# Lecture Notes in Computer Science 14831

Founding Editors

Gerhard Goos
Juris Hartmanis

The series Lecture Notes in Computer Science (LNCS), including its subseries Lecture Notes in Artificial Intelligence (LNAI) and Lecture Notes in Bioinformatics (LNBI), has established itself as a medium for the publication of new developments in computer science and information technology research, teaching, and education.

LNCS enjoys close cooperation with the computer science R & D community, the series counts many renowned academics among its volume editors and paper authors, and collaborates with prestigious societies. Its mission is to serve this international community by providing an invaluable service, mainly focused on the publication of conference and workshop proceedings and postproceedings. LNCS commenced publication in 1973.

Meiko Jensen · Cédric Lauradoux ·
Kai Rannenberg
Editors

# Privacy Technologies and Policy

12th Annual Privacy Forum, APF 2024
Karlstad, Sweden, September 4–5, 2024
Proceedings

 Springer

*Editors*
Meiko Jensen 🆔
Karlstad University
Karlstad, Sweden

Cédric Lauradoux 🆔
Université Grenoble Alpes, Inria
Montbonnot-Saint-Martin, France

Kai Rannenberg
Goethe University Frankfurt
Frankfurt am Main, Germany

ISSN 0302-9743      ISSN 1611-3349 (electronic)
Lecture Notes in Computer Science
ISBN 978-3-031-68023-6      ISBN 978-3-031-68024-3 (eBook)
https://doi.org/10.1007/978-3-031-68024-3

# Preface

The 2024 edition of the Annual Privacy Forum (APF 2024) was co-organized by the European Union Agency for Cybersecurity (ENISA), DG Connect of the European Commission, and Karlstad University. The conference was hosted in Karlstad, Sweden, in September 2024. This conference, already in its 12th edition, was established as an opportunity to bring together key communities, namely policy, academia, and industry, in the broader area of privacy and data protection while focusing on privacy-related application areas. Like in the previous edition, a large focus of the 2024 conference was on the General Data Protection Regulation (GDPR) and the emerging legislation around the European Data Spaces and Artificial Intelligence. There were 60 submissions in response to the APF call for papers. Each paper was peer-reviewed by at least three members of the international Program Committee (PC). On the basis of significance, novelty, and scientific quality, twelve papers were selected (a 20% acceptance rate) and are compiled in this volume.

We wish to thank the members of the Program Committee for devoting their time to reviewing the submitted papers and providing constructive feedback, the authors, whose papers make up the bulk of the content of this conference, and the attendees, whose interest in the conference is the main driver for its organization.

September 2024

Prokopios Drogkaris
Meiko Jensen
Cédric Lauradoux
Kai Rannenberg

# Organization

## General Chairs

Meiko Jensen          Karlstad University, Sweden
Cédric Lauradoux      Inria, France
Kai Rannenberg        Goethe University Frankfurt, Germany

## Program Committee

Isabel Barberá                Rhite, The Netherlands
Roman Bieda                   Kozminski University, Poland
Athena Bourka                 ENISA, Greece
Robert Cronk                  Foryte Web Services Inc., USA
Giuseppe D'Acquisto           Garante per la protezione dei dati personali, Italy
Silvia De Conca               VU Amsterdam, The Netherlands
José M. Del Álamo             Universidad Politécnica de Madrid, Spain
Matteo Dell'Amico             University of Genoa, Italy
Vasiliki Diamantopoulou       University of the Aegean, Greece
Prokopios Drogkaris           ENISA, Greece
Ana Ferreira                  University of Porto, Portugal
Simone Fischer-Hübner         Karlstad University, Sweden
Michael Friedewald            Fraunhofer ISI, Germany
Lothar Fritsch                Oslo Metropolitan University, Norway
Olga Gkotsopoulou             Vrije Universiteit Brussel, Belgium
Elias Grünewald               Technische Universität Berlin, Germany
Nils Gruschka                 University of Oslo, Norway
Agnieszka Gryszczyńska        Cardinal Stefan Wyszynski University in Warsaw
                              (UKSW), Poland
Marit Hansen                  Unabhängiges Landeszentrum für Datenschutz
                              Schleswig-Holstein, Germany
Jaap-Henk Hoepman             Radboud University, The Netherlands
Kristina Irion                University of Amsterdam, The Netherlands
Leonardo Iwaya                Karlstad University, Sweden
Meiko Jensen                  Karlstad University, Sweden
Christos Kalloniatis          University of the Aegean, Greece
Irene Kamara                  Tilburg University, The Netherlands
Liina Kamm                    Cybernetica AS, Estonia

| | |
|---|---|
| Sokratis Katsikas | Norwegian University of Science and Technology, Norway |
| Cedric Lauradoux | Inria, France |
| Konstantinos Limniotis | Hellenic Data Protection Authority, Greece |
| Tanel Mällo | Cybernetica AS, Estonia |
| Teresa Martínez Sánchez | Spanish Data Protection Authority, Spain |
| Victor Morel | Chalmers University of Technology, Sweden |
| Maria Owczarek | Urzad Ochrony Danych Osobowych, Poland |
| Frank Pallas | Paris Lodron University of Salzburg, Austria |
| Daniela Pöhn | Universität der Bundeswehr München, Germany |
| Maria Grazia Porcedda | Trinity College Dublin, Ireland |
| Kai Rannenberg | Goethe University Frankfurt, Germany |
| Delphine Reinhardt | University of Göttingen, Germany |
| Nikita Samarin | University of California, Berkeley, USA |
| Ina Schiering | Ostfalia University of Applied Sciences, Germany |
| Stefan Schiffner | Berufliche Hochschule Hamburg BHH University of Applied Sciences, Germany |
| Jan Tolsdorf | Bonn-Rhein-Sieg University of Applied Sciences, Germany |
| Tom Van Cutsem | KU Leuven, Belgium |
| Isabel Wagner | University of Basel, Switzerland |
| Christian Zimmermann | Robert Bosch GmbH, Germany |

## Additional Reviewers

Jörg Aßmann
Valerie Fetzer
Lindrit Kqiku

# Contents

# Implications of Age Assurance on Privacy and Data Protection: A Systematic Threat Model
### Research Paper

Marta Beltrán$^{(\boxtimes)}$ ⓘ and Luis de Salvador

Agencia Española de Protección de Datos (AEPD), Madrid, Spain
{mbeltran,lsalvadorc}@aepd.es
https://www.aepd.es/en

**Abstract.** In today's digital world, children are encouraged to develop a significant part of their daily lives by online means. Age assurance solutions have become essential tools to ensure the protection of their fundamental rights. This is reflected in different regulatory frameworks, strategies, codes, and recommendations concerning children's protection and a safer Internet. These solutions estimate or verify users' ages, allowing for the establishment of age restrictions for content, services, and goods. However, the intersection of safety and privacy within this field poses significant challenges, not only for children but for all Internet users.

This research explores the privacy and data protection implications of age assurance, analyzing existing solutions and proposing a comprehensive privacy threat model. The threat model developed in this paper can be widely applied to improve the design and implementation of current solutions, policy and guidelines formulation, and awareness initiatives about age assurance challenges. By balancing adequate age assurance with robust and compliant data protection, we can create a digital environment for everyone that guarantees the protection of fundamental rights.

**Keywords:** Age assurance · Children protection · Data protection · Privacy risk management · Threat model

## 1 Introduction

In an increasingly digital world, where children and young users actively engage with online content and services, ensuring their safety becomes paramount. Policymakers have recognised the significance of age assurance mechanisms within this context in different regulatory frameworks [1,3,4] and design codes [2,18]. These solutions, designed to estimate or verify users' age, are devoted to safeguarding children from harmful content, age-inappropriate interactions, exploitative practices, or the purchase of age-restricted goods, to mention some examples [24].

M. Jensen et al. (Eds.): APF 2024, LNCS 14831, pp. 1–22, 2024.
https://doi.org/10.1007/978-3-031-68024-3_1

Age assurance solutions play an essential role in determining the age of users, but they also raise concerns about privacy and data protection [28]. These solutions often require collecting personal data such as birth dates, identification documents, or biometric information, which can pose risks when processing, storing, or sharing this data. There is often a lack of transparency regarding user age data processing, with many providers relying on third-party age assurance services. This can introduce additional privacy threats, as users may not know how their data flows beyond the platform they interact with. Moreover, when combined with other identifiers such as IP addresses or browsing history, age assurance data can potentially identify users, revealing more information about them than intended.

In addition, age assurance solutions should focus on estimating or verifying age. However, over time, some of them are expanding their functionality to collect additional data (e.g. identity, location, interests), generating parallel identity frameworks with new and different purposes different from age assurance. In these cases, age assurance solutions may inadvertently create profiles based on user data. These profiles can be used for targeted advertising, behavioural analysis, or discriminatory practices [13].

Age assurance solutions must manage a careful balance between safeguarding children and respecting the rights of all individuals. To effectively mitigate these risks, strong privacy-enhancing measures are necessary. However, it is important to note that there is no comprehensive list of all potential privacy threats arising from data processing for age assurance purposes.

This research paper contributes significantly to the field by 1) Analysing existing age assurance solutions, bringing to light their common aspects and identifying the most prevalent architectures and their privacy implications. 2) Proposing a comprehensive and systematic threat model to identify privacy threats specific to age assurance systems. 3) Discussing the produced model to offer practical recommendations to enhance privacy while maintaining adequate age assurance.

The rest of this paper is organised as follows: Sect. 2 summarises the related work on age assurance and threat modelling and outlines the motivation of this research. Section 3 describes our research method. Section 4 analyses the most extended architectures used to solve age assurance. Section 5 describes the produced privacy threat model, while Sect. 6 discusses the identified threats and provides initial recommendations. Finally, Sect. 7 summarises our main conclusions.

## 2   Related Work and Motivation

### 2.1   On Age Assurance

Age assurance is a term adopted to describe different methods used to determine users' age on online platforms, applications, and services [5]. Age estimation and age verification are two critical components of age assurance, each having a specific function. Age estimation uses probabilistic methods based on biometry (often voice or facial analysis), capacity testing, behaviour patterns, and language use to guess a user's age. Age verification, conversely, guarantees age by using an existing digital identity or providing a document, such as a passport or other government-issued ID, to confirm an individual's age. A subcategory within verification is composed of the methods inferring the age or age range due to the possession of a particular document or certificate, such as a bank card or digital credential.

Previous work has identified the main challenges concerning age assurance:

- Accuracy: False positives and negatives can occur, leading to incorrect age assurance.
- Lack of Standards: The absence of a unified standard and the adoption of diverse strategies and approaches create a fragmented landscape, presenting challenges for the industry in integrating their platforms, applications and services with different alternatives.
- Cost: Given the fragmentation mentioned above, implementing age assurance systems may be expensive, especially for small and medium-sized providers.
- User Experience: Assurance processes might drive users away from the platform due to a negative experience (usability issues, lack of accessibility, etc.).
- Universality and Inclusiveness: There are limited options for assuring users' age online. Each method has its advantages and disadvantages, but covering all domains' and users' requirements is challenging without causing exclusion.
- Compliance: Providers must ensure compliance with different regulations (children protection, online safety, digital services, audiovisual contents, data protection, etc.) and keep up with changing and heterogeneous legal requirements.
- Security: Users trying to access unauthorized content or services can circumvent or work around age assurance systems. Malicious actors may also be interested in attacking these systems to steal personal data, commit fraud, or carry out some impersonation.
- Privacy and Data Protection: Age assurance systems often require users to provide personal data, which raises privacy concerns.

Research efforts have been focused on exploring ways of overcoming some of these challenges [11, 20]. However, most of the works focus on discussing whether it is the most appropriate approach in different use cases [7, 8, 14, 22, 29] or on ethical and public policy aspects [9, 25, 32].

Furthermore, the British Standards Institute (BSI) and the Digital Policy Alliance developed a code of practice for online Age verification service providers. This code, called PAS 1296:2018 [10], applies to providers who must perform age assurance processes to determine whether or not a citizen can access age-restricted goods, content or services.

The International Standards Organisation (ISO) is currently working on developing ISO/IEC 27566 [19]. One of the main drivers for this effort is the growing agreement that a straightforward method to explain the levels of confidence attained by different assurance components would benefit service providers, relying parties, and regulators.

Finally, it is worth mentioning two European initiatives. The first, the euCONSENT project [15], working on proposing operation extensions to the eIDAS (electronic IDentification, Authentication and trust Services) infrastructure to deliver pan-European, open-system, secure and certified interoperable age verification and parental consent to access Information Society Services. The second, the Task Force on Age Verification under the Digital Services Act, set up in 2024 to progress towards a harmonised EU approach to age verification [12].

## 2.2   On Threat Modelling

Threat modelling is a systematic process used to identify, understand and communicate threats and mitigations while protecting something valuable. It is a structured representation of all the information that affects the safety, security or privacy of a wide range of assets, including software and applications, systems, networks or business processes [31].

While traditional threat modelling concerns safety and security threats, privacy threat modelling explicitly addresses risks related to collection, recording, organisation, structuring, storage, adaptation or alteration, retrieval, consultation, use, disclosure by transmission, dissemination or otherwise making available, alignment or combination, restriction, erasure or destruction of personal data.

Privacy threat modelling has been applied in various domains. The LIND-DUN methodology is the most widely used in the field of privacy engineering [30]. It can anticipate and prevent data processing that may lead to privacy risks or analyse privacy requirements. However, it has also been used as a mnemonic for a brainstorming-style exercise rather than a means to elicit privacy threats systematically [16,23,26]. Furthermore, since a PIA (Privacy Impact Assessment) or a DPIA (Data Protection Impact Assessment) implies a risk assessment which typically follows a step-by-step process of risk identification and risk mitigation, a LINDDUN-based privacy threat model can also be a systematic and traceable starting point to conduct this kind of exercises [17].

## 2.3   Motivation

To date, no systematic threat model has been proposed for identifying the privacy and data protection threats that arise when using age assurance systems. This model would provide critical insights into the adequacy of existing solutions in terms of both data protection and regulatory compliance, as well as their alignment with the rights and freedoms of citizens. In the absence of such a model, making recommendations to relevant stakeholders, including providers, third parties, regulators, control authorities, and users, has been challenging. Adopting a systematic threat model for data protection in age assurance systems promises to address these challenges by providing a comprehensive framework for evaluating existing solutions and identifying opportunities for improvement.

## 3    Method

This research proposes a method with two main phases to address the identified research gap.

1. Architectures analysis: The first phase involves reviewing the age assurance literature and analysing the proposed architectures (see Sect. 4). Searches of standards, scientific papers, commercial products, and patents need to be carried out using the keywords age assurance, age estimation, and age verification. The aim is not to produce a survey of the different solutions and methods but to understand the targeted systems and their common aspects: processes, involved agents, information flows, data formats, etc. These common aspects in the analysed architectures can be used as the input for the second phase.
2. Threat model: This study's second phase involves the threat modelling process, as outlined in Sect. 5. To accomplish this, the LINDDUN methodology has been selected, a well-established and privacy-focused threat modelling approach that has been empirically evaluated. We have adopted the LINDDUN methodology in two critical aspects of our work. Firstly, a Data Flow Diagram (DFD) has been used as a graphical tool to model the system under analysis and guide the threat identification and analysis process. The LINDDUN catalogues and body of knowledge have been used to aid in this process [21]. Secondly, we have analysed threats across various categories, some consistent with the LINDDUN approach. In contrast, others have been modified or added to align with the primary objective of this research, which is data protection, regulatory compliance, and the protection of individuals' rights and freedoms. The LINDDUN methodology is an acronym for Linking, Identifying, Non-repudiation, Detecting, Data disclosure, Unawareness, and Non-compliance. Some categories have been retained, while others have been removed or added as necessary. Specifically, we have eliminated the Non-compliance category, from the regulatory point of view, as data processing that does not comply with regulations cannot be implemented. Additionally, we have added three new categories: Inaccuracy, Exclusion, and Data breaches. These categories better reflect the focus of our research and enable us to more effectively identify and address potential threats to data protection and regulatory compliance. The considered categories, LIINE3DU, are then as follows:
    - Linking: This threat involves associating different data items or user actions to learn more about a data subject or group.
    - Identifying: This threat involves learning the identity of a data subject directly (through leaks, for example) or indirectly (through deduction or inference, for example).
    - Inaccuracy: This refers to using obsolete, wrong, incomplete or low-quality data that may lead to incorrect decisions or actions, potentially causing inconvenience or even harm to the data subject.
    - Non-repudiation: This refers to the ability to attribute a claim to the data subject (something they know, they are, they do, etc.).

- Exclusion: This threat involves unintentionally or deliberately failing to adequately serve a data subject, hindering their participation or involvement in physical or digital life.
- Detecting: This threat involves deducing the existence of data items or user actions through observation.
- Data Breach: This threat involves destruction, loss, alteration, unauthorised disclosure of, or access to, personal data by mistakes, malicious insiders or cyberattacks.
- Data Disclosure: This threat involves excessively collecting, storing, processing, or sharing personal data.
- Unawareness and Unintervenability: This threat involves insufficiently informing, involving, or empowering data subjects in the processing of their personal data.

The last two categories are intrinsically linked to an organisation's risk management processes and to the data processing design, and entail non-compliance with data protection regulatory frameworks. Conversely, the remaining proposed categories may directly impact individual rights and freedoms from a broader perspective.

## 4   Architectures Analysis

The market for age assurance is changing with the introduction of new products and services. Additionally, regulatory actions for online protection of children and data privacy are shaping the market.

Regarding age assurance, self-declaration is no longer considered a reliable solution in almost all use cases. In light of this, three different architectural approaches have emerged. The first approach involves direct interaction between the user and the provider, who is responsible for ensuring their age without the involvement of specialised providers. The other two approaches are "tokenised", with the provider delegating age assurance to a specialised third-party provider capable of returning the necessary information about the user in a date of birth, an age, an age range or an age threshold fulfilment token.

When a user wants to access age-restricted content or services, the data flow begins with an access request. During this process, participating agents store data based on their roles and level of access. This can include user data, metadata, logs, and the age assurance result.

### 4.1   Age Assurance Solved by the Content or Service Provider

In this first architecture, the steps followed to solve age assurance are:

1. Age Assurance Request: Once the user attempts to access age-restricted content or services, the system triggers the age assurance process. The system prompts the user to provide information or evidence concerning their age.

2. Data Collection: Depending on whether the system uses age estimation or age verification, the data collected will vary. For Age estimation the system may collect data such as biometry (voice samples or face images), behavior patterns, language use, or other indirect indicators of age. In the case of Age verification, the user is asked to provide a reliable proof of age. This could be in the form of a government-issued ID, a passport, an electronic certificate or credential, credit card information, or another form of age verification accepted by the system.

3. Age Check: The system processes the collected data in-house. For Age estimation the system uses statistical methods, machine learning or artificial intelligence models to estimate the user's age based on the collected data. For Age verification the system checks the provided proof, for example, the date of birth on a government-issued ID, the age on an electronic certificate or credential or the confirmation that a credit card is valid and belongs to the user.

4. Result Communication: The system then communicates the result of the age assurance operation to the user. If the user's age is successfully assured, they are granted access to the age-restricted content or service. If the age assurance process fails, the user is denied access and may be given a reason for the denial.

## 4.2   Age Assurance Solved by a Third-Party Provider as a Proxy

In this second architecture a new agent is necessary to solve age assurance: a third-party specialised provider. The provider offering contents or services with age-restrictions adopt the role of Relying party, delegating the responsibility on this third-party provider. The steps followed to solve age assurance are:

1. Age Assurance Request: Once the user attempts to access age-restricted content or services, the system triggers the age assurance process. Instead of directly interacting with the user, the system redirects them to a third-party age assurance provider.

2. Data Collection at the Third-Party Provider: The user is then asked to provide the necessary information or evidence for age estimation or verification to this third-party provider.

3. Age Check by the Third-Party Provider: The system processes the collected data. For Age estimation the system uses statistical methods, machine learning or artificial intelligence models to estimate the user's age based on the collected data. For Age verification the system checks the provided proof.

4. Result Communication: The third-party provider then communicates the result of the age assurance process to the original service provider. The service provider then communicates the result to the user. If the user's age is successfully assured, they are granted access to the age-restricted content or service. If the age assurance process fails, the user is denied access and may be given a reason for the denial.

### 4.3   Age Assurance Solved by a Third-Party Provider as an Autonomous Agent

In this last architecture, a third-party specialised provider is used again. However, it does not act as a proxy between the user and the content or service provider in this case. It only produces the age token the user must present at the content or service provider to ensure their age. This architecture could be understood as a particular case of architecture 1, where the data collected at step 2 is this age token produced by a reliable external agent capable of estimating or verifying the user's age. In this research, we prefer to analyse it as a different architecture to consider its peculiarities in terms of data protection.

The steps followed to solve age assurance in this case are:

1. Age Assurance Request: Once the user attempts to access age-restricted content or services, the system triggers the age assurance process. Instead of directly interacting with the user, the system asks them for a electronic capability assured by a third-party age assurance provider.
2. Data Collection at the Third-Party Provider: The user is then asked to provide the necessary information or evidence for age estimation or verification to this third-party provider.
3. Age Check by the Third-Party Provider: The system processes the collected data. For Age estimation the system uses statistical methods, machine learning or artificial intelligence models to estimate the user's age based on the collected data. For Age verification the system checks the provided proof. The third-party provider generates an electronic capability (signed certificate or unforgeable token), usually only if the user's age is successfully assured. This capability is sent to the user so the user can present it at the original service provider. If the age assurance process fails, the user is usually denied the age token and may be given a reason for the denial.
4. Result Communication: If the user's age is successfully assured, they are granted access to the age-restricted content or service. If the age assurance process fails, the user is denied access and may be given a reason for the denial.

Steps 2 and 3 of this process may or may not be synchronized with access to the content or service provider. This means that the user can request the age token at the moment when they need to assure their age to a content or service provider, or they can do it offline, at a prior time, and store one or more age tokens in a securely protected repository, trusted for future use. In this paper, only the synchronized approach is considered (when the age token is requested when needed) because it is the one that has been found already implemented in real solutions that can be analysed and tested for the proposed threat model.

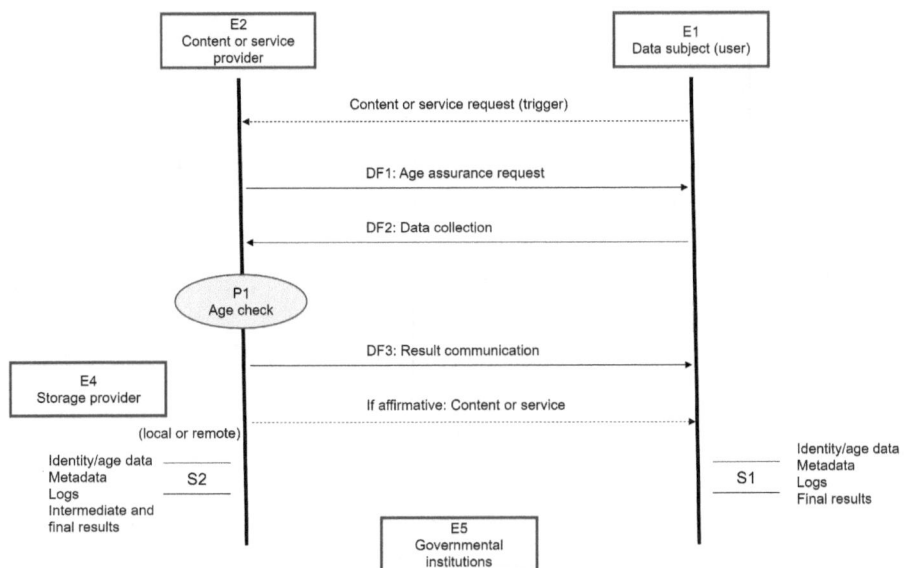

**Fig. 1.** Architecture 1 (DFD): Age assurance solved by the content or service provider.

## 5   Threat Model

The Data Flow Diagrams (DFD) produced to conduct the threat modelling process come directly from the description provided in the previous section for the different architectures (Figs. 1, 2 and 3).

Entity E1 is the subject whose age is to be assured. Entity E2 is a content or service provider offering content or services with age restrictions, while entity E3 is the third party specialised in age assurance. E2 and E3 have its own storage, which can be local or remote. For this reason, E4 is a Storage as a Service provider. And E5 are governmental institutions such as judges and courts, security forces and bodies, etc.

All data flows from DF1 to DF7 in the different architectures are those that have been identified in this research as essential when performing age assurance: Age assurance request, Data collection and Result communication, but also Service request and Service result when a specialised third-party provider is involved. These are the minimum data contents of all these essential flows:

– Age assurance request: DF1 in architecture 1, DF2 in architecture 2 and DF3 in architecture 3, this flow includes at least the age threshold that must be exceeded to gain access to the requested content or service.
– Data collection: DF2 in architecture 2, DF3 in architecture 2 and DF4 in architecture 3, this flow includes personal data. What specific data depends on the method used to perform age assurance, estimation or verification, and the specific approach (biometry, date of birth, etc.).

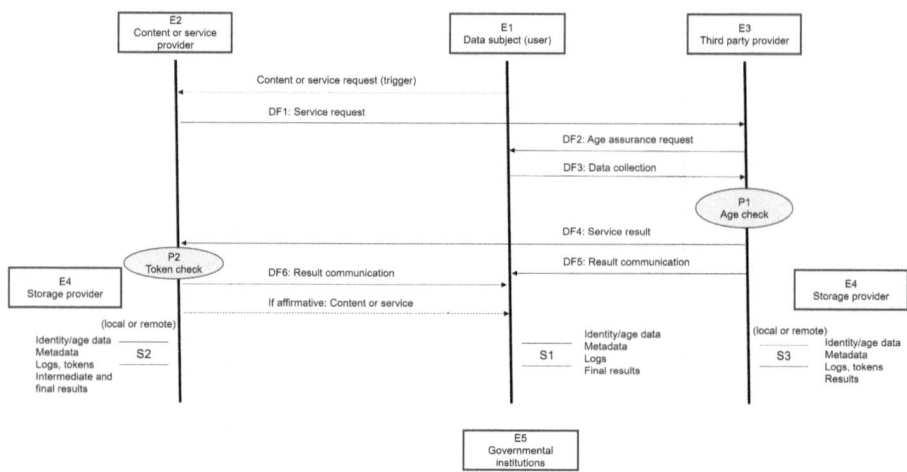

**Fig. 2.** Architecture 2 (DFD): Age assurance solved by a third-party provider as a proxy.

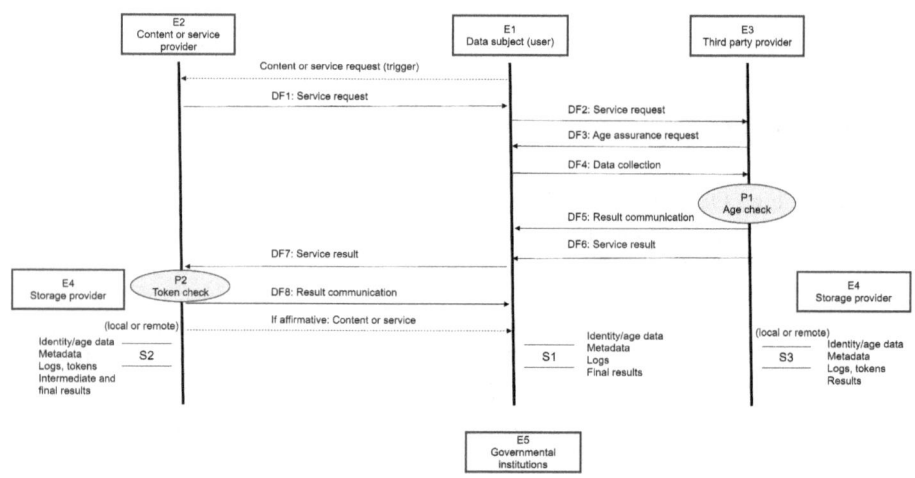

**Fig. 3.** Architecture 3 (DFD): Age assurance solved by a third-party provider as an autonomous agent (synchronously).

- Result communication: DF3 in architecture 1, DF5 and/or DF6 in architecture 2, and DF5 and/or DF8 in architecture 3. This flow includes a passed (age above threshold)/not passed (age not above threshold, no consent, error) and a motivation for this result.
- Service request: DF1 in architecture 2 and 3, DF2 in architecture 3. This flow includes, at least, information about this specific age assurance process, such as a transaction number or a nonce. It may also include the age threshold.

– Service result: DF4 in architecture 2, DF6 and DF7 in architecture 3. This flow includes a token with information about the user's date of birth, age ("24 years old", for example), the user's age range ("+18", for example) or the fulfilment of the age threshold (passed/not passed). This token is usually signed to allow the verification of its origin and usually includes an expiration time.

Finally, two different processes have been identified. P1 is the Age check, a simple comparison between the user's age and the age threshold. P2 is the Token check, the interpretation and processing of the information received in the token and the verification of its validity (it has not expired, the origin is trusted, etc.). Different agents can execute these processes depending on the specific analysed architecture.

These DFDs have been produced considering the following assumptions:

1. The developed threat model considers that age assurance processes are data processing by themselves.
2. The produced threat model is focused exclusively on the rights and freedoms of the subject undergoing the age assurance process.
3. All elements of the DFD that are the responsibility of the content or service provider and the third parties with which it works are sufficiently protected against malicious insiders and cyberattacks, as well as the communications between them. There is a trusted boundary that groups the essential elements for age assurance.
4. The processes within the DFD are appropriately implemented and perform the function for which they are designed.
5. The elements appearing in the DFD cannot be impersonated except for the subject undergoing identity verification.

In this way, threats related to security are not modelled (a different and complementary modelling process would be needed) but exclusively to privacy and data protection. Under all these assumptions and after analysing 12 different solutions (8 already in production and 4 in a prototype phase), the following main threats have been identified.

## 5.1 Linking

**Threat 1: Unique Identifier.** A threat agent is able to learn more than expected about a data subject and their actions by associating this data subject to a unique identifier: this identifier enables the linking of all accesses/requests to the same individual (without identifying this data subject).

– Architecture 1: E2 is able to materialise this threat through the DF2, containing personal data, and through S2, if it stores some of this personal data. E2 is able to link collected data, for example, an email address, a user ID or a face template, with the user's content requests.

- Architecture 2: E2 is able to materialise this threat through the DF4, containing the token data, if it contains unique identifiers. Moreover, through S2 if it stores some of this token data. E2 is able to link token data, such as an email address, a user ID, or a face template, with the user's content requests. E3 can also materialise this threat through the DF3, which contains personal data, and S3 if it stores some of this personal data. E3 is able to link collected data, for example, an email address, a user ID or a face template, with the service requests coming from the content or service provider (E2).
- Architecture 3: E2 is able to materialise this threat through the DF7, containing the token data, if it contains unique identifiers. Furthermore, through S2 if it stores some of this token data. E2 is able to link token data, such as an email address, a user ID, or a face template, with the user's content requests. E3 is able to materialise this threat through the DF4, which contains personal data, and S3, if it stores some of this personal data. E3 is able to link collected data, for example, an email address, a user ID or a face template, with the service requests coming from the user (E1).

It is worth noting that the privileged position of the third-party provider in architectures 2 and 3 allows it to trace one user's interactions with different content and service providers. This means that all online activities related to adult content or sites could be associated with this specific data subject.

Any entity E4 could also materialise this threat if any storage elements are remote and consumed as a service from an external provider.

**Threat 2: Linkable Data Through Combination, Profiling or Inference.**
A threat agent is able to learn more than expected about a data subject and their actions by associating this data subject to all their accesses/requests (without a unique identifier and without identifying this data subject). Access requests and service requests usually contain attributes that, when combined, are unique to a data subject. The more data is collected (volume and variety) and the more detailed it is, the easier it is to find unique patterns for linking.

- Architecture 1: E2 is able to materialise this threat through the DF2 and S2, which may contain a variety of metadata and logs. For example, E2 may link different requests to the same user tracing their IP address, their location, combining data such as their date of birth and their postal code, performing browser fingerprinting during their interactions, analysing timing patterns or analysing behaviour or language.
- Architecture 2: E2 is able to materialise this threat through the DF4, containing the token data, depending on its content (and through S2). E2 is able to link token data, such as IP address, location, date of birth, and postal code, with the user's content requests. E3 can also materialise this threat through the DF3, which contains personal data, and through S3. E3 is able to link the above-mentioned "quasi-identifiers" or users' profiles collected or built during user interactions, with service requests coming from the content or service provider (E2).

– Architecture 3: E2 is able to materialise this threat through the DF7, containing the token data and through S2. E3 is able to materialise this threat through the DF4, containing personal data, and through S3. E3 is able to link "quasi-identifiers" or user's profiles collected or built during user's interactions with the service requests coming from the user (E1).

## 5.2    Identifying

**Threat 3: Identified Data.** A threat agent is able to collect and process the data subject identity within the age assurance process.

– Architecture 1: E2 is able to materialise this threat through the DF2 and S2, which may contain a variety of metadata and logs. For example, E2 requires a registration that includes a full name. Alternatively, when the user ID within the platform is based on a combination of first and last names.
– Architecture 2: E2 is able to materialise this threat through the DF4, containing the token data, depending on its content (and through S2). For example, name and surname. E3 can also materialise this threat through the DF3, which contains personal data, and through S3.
– Architecture 3: E2 is able to materialise this threat through the DF7, containing the token data and through S2. E3 is able to materialise this threat through the DF4, which contains personal data, and S3.

**Threat 4: Identifiable Data.** A threat agent is able to learn or infer the data subject's identity within the age assurance process. For example, reversing a pseudonym, using a reverse image search tool to identify the data subject from a face image, etc. Different attributes or combinations of attributes are unique to a specific data subject and enable their identification. The scenarios are the same as in threat 3, but the threat agent does not collect/process the identity directly but instead has to perform some operation to derive it. There is a particular case when an entity E5 (usually, it has to be governmental to have the capacity to materialize this threat) is the only one that can identify the data subject, requesting the necessary data from entities 2 and 3 to do so. Each of these entities, separately, cannot identify the user with the data they handle. However, E5, collecting the data processed by both (for example, with a court order), can infer the data subject identity.

## 5.3    Inaccuracy

**Threat 5: Error.** Age assurance is performed by processing obsolete, wrong, incomplete or low-quality data. The threat materialises unintentionally. For example, the identity document with which the age verification is carried out has expired or the data subject has turned years old and their data still needs to be updated. In the case of age estimation, the data used to make the age prediction is not of sufficient quality. In a cold start, for example, with a new user about whom the provider barely has any information or with an image of a face taken in bad lighting. There are also errors inherent to estimation methods since, by their very nature, they cannot guarantee 100% accuracy (methods performance, potential bias or discrimination, etc.). This threat is not only about direct data collection, it can also be materialised with expired or deficient tokens in architectures 2 and 3. The elements of the DFD involved in this threat are the Data collection flows (DF2, DF3, and DF4 in architectural designs 1, 2, and 3, respectively) and the processes P1 and P2, Age check and Token check.

**Threat 6: Circumvention and Evasion.** Age assurance is performed processing obsolete, wrong, incomplete or low-quality data, intentionally because a user wants to pretend they have a different age from their real one. User's binding is essential in verification methods, for example, since one user could use another's document, certificate or credential if proper checks are not performed. Alternatively, because the E2 or E3 entities have an interest (commercial, directly or indirectly) in ensuring ages differ from their real ones. For example, to allow more users to access the platform, to monetize a more significant number of positive results, etc. The elements of the DFD involved in this threat are again the Data collection flows (DF2, DF3, and DF4 in architectures 1, 2, and 3, respectively) and the processes P1 and P2.

### 5.4   Non-repudiation

**Threat 6: Attributable Data Evidence.** A threat agent can use data to prevent the data subject from denying an access/request. Threats 1, 2, 3 and 4 may magnify the impact of this one because a group of different actions can be attributed to the same data subject (by materialising Linking) even knowing who this data subject is (by materialising Identifying).

- Architecture 1: E2 is able to materialise this threat through S2, which may contain a variety of data, metadata and logs.
- Architectures 2 and 3: E2 is able to materialise this threat through S2 and E3 through S3. Digitally signed tokens are an additional source of attributable data evidence that may prevent the data subject or E3 from denying their involvement in a transaction later.

There is a particular case when entity E5 is the only one that can attribute an action to the data subject, requesting the necessary data from entities 2 and 3 to do so. Even requesting the contents of S1 in the user's device, for example, the Result communication logs that could be combined with browser history to substantiate claims about their online activities.

### 5.5   Exclusion

**Threat 7: Limited Access.** A data subject is marginalised, unable to complete age assurance processes, which hinders their participation in different online activities and restricts their autonomy because they have restrictions to use the required elements or technologies. For example, a user does not have a smartphone with a high-quality camera or an NFC reader, or is not in possession of the document or certificate requested to verify their age.

**Threat 8: Limited Literacy or Confidence.** In this case the data subject is marginalised because they have a low-proficiency level, lack of knowledge or lack of confidence in the provided methods. For example, the user does not have the capability to create an account in a specific platform, or does not want to use a governmental solution.

## 5.6  Detecting

**Threat 9: Observed Communications.** The access/request of a data subject or an attribute of this subject can be deduced through the observation of communication within the DFD. For example, Service request data flows at E3 (architectures 2 and 3) imply that users are trying to access age-restricted content or services. If threats 1, 2, 3 or 4 have been materialised, this user is also a specific one. Observing Service result flows (architectures 2 and 3) and Result in communications, in general, can also enable detection, even inferring if a user is a child or their age range.

**Threat 10: System Responses.** The access/request of a data subject or an attribute of this subject can be deduced by examining system responses. For example, E2 or E3 can infer if a user is a child or their age range with the results obtained from P1 and P2. Logs, error messages, metadata, etc., stored in S1, S2 and S3 also allow detection.

## 5.7  Data Breach

**Threat 11: Breach of Stored Data.** Even when proper cybersecurity risk management procedures have been implemented, data breaches affecting information that resides in databases, files or other storage systems may be inevitable. They can affect S1, S2, and S3 and all the personal data stored in these repositories (also in logs, error messages, metadata, etc.). Threats 1, 2, 3 and 4 may magnify the impact of this one because the external threat agent materialising this threat can attribute the leaked data to a specific data subject.

**Threat 12: Breach of Data in Transit.** In this case the breach affects data in motion being transferred between two DFD elements over networks. The elements of the DFD involved in this threat are the Data collection flows (DF2, DF3, and DF4 in architectures 1, 2, and 3, respectively) but also the rest of the flows containing personal data, for example, Service result flows containing tokens in architectures 2 and 3. Threats 1, 2, 3 and 4 may magnify the impact of this one because the external threat agent materialising this threat can attribute the leaked data to a specific data subject.

**Threat 13: Breach of Data Being Processed.** The data breach affects data being actively manipulated by applications, scripts or services. Sometimes, it is during processing that the data has the most value, because it is usually necessary to decrypt it in order to work with it. The elements of the DFD involved in this threat are P1 ans P2 when they process personal data.

## 5.8    Data Disclosure

**Threat 14: Unnecessary Data Types or Volume.** The data minimisation principle is not fulfilled, and personal data is not adequate, relevant, or limited to what is necessary concerning the age assurance process. E2 and E3 can materialise this threat through the Data collection flows (DF2, DF3, and DF4 in architectures 1, 2, and 3, respectively). For example, a provider may ask the user for more personal data (name, surname, address, email, postal code, etc.) than functionally required by the solution, which only requires data concerning age. In fact, in most cases it is not necessary to know the date of birth, age or age range, simply if the user's age exceeds a certain threshold. It must be taken into account that the data is sensitive, not only because it could allow minors to be located but also other people from vulnerable groups such as the old adults. Another example is providers that use more than one age assurance method for the same user/process, when with only one it is possible to check whether the user can access the requested content or service. In other cases, personal data is collected more frequently than functionally needed, for example, when the requested content is for all audiences instead of age-restricted. Furthermore, when personal data is collected of more data subjects than functionally needed, for example, children's data when only the adults' age should be ensured. Special mention deserves the cases in which special categories of data are processed without taking into account that their processing is prohibited unless any of the assumptions included in the regulation occur. This is the case, for example, of age estimation when it is based on biometric data enabling the identification of a natural person.

**Threat 15: Unnecessary Processing.** The purpose limitation principle is not fulfilled, and the collected personal data are processed in a manner incompatible with the initially specified, explicit and legitimate purposes concerning age assurance. E2 and E3 can materialise this threat through P1, P2, or other new processes different from those required to conduct age assurance. There are many cases where personal data may be further treated, analysed, and enriched in an unnecessary way to achieve the solution's functionality. For example, providers relying on age estimation may use users' face templates to analyse facial expressions or to recognise emotions, users' behaviour datasets to profile them or analyse their habits, etc. Retaining users' personal data long after they have been through the age assurance process may also be problematic (in S2 and S3), not fulfilling the storage limitation principle. Some providers, for example, use the data to test different methods or to train machine learning or artificial intelligence models (research&development purposes).

**Threat 16: Unnecessary Exposure.** In this case, personal data are made accessible to more parties than functionally necessary. E2 and E3 can materialise this threat through P1, P2, S2, and S3 or create new data flows different from those required to perform the age assurance process. For example, we have found providers propagating data to third-party providers, offering them storage services (E4) but also AI services (for age estimation), liveness tests in real-time (when the veracity of a physical document needs to be verified), additional background checks (banks or telecom operators are sometimes included in age assurance flows because they know their customers and can help to verify some of their attributes), etc. This threat also materialises when E2 or E3 share data with advertisers or use tracking and analytics services, to mention only some additional examples. Alternatively, when they publish research datasets.

### 5.9 Unawareness and Unintervenability

**Threat 17: Lack of Information.** In this case the lawfulness, fairness and transparency principle is not fulfilled and the data subject is not aware about the collection, processing, storage, or disclosure of their personal data. This threat can be materialised by E2 and E3. For example, long and incomprehensible privacy notices leading the data subject to not be aware of the collected data and metadata, the processing purposes, the third parties with whom their data will be shared, retention periods, etc. We have also found that the default settings of different solutions may not adhere to the highest standards of privacy and may employ dark patterns that can influence users to inadvertently share more data or consent to terms that they do not fully comprehend.

**Threat 18: Lack of Control.** In this case the data subject is not capable of exercising their data rights (the right of access, the right to rectification, the right to erasure, the right to restrict processing, the right to data portability, the right to object and the right not to be subject to a decision based solely on automated processing). This threat can be materialised by E2 and E3.

## 6 Discussion

Age-restricted content and services providers and third-party specialised providers must assess the potential impacts for privacy and data protection associated with their age assurance solution and their design and implementation decisions. This process should fulfil all the fundamental rights protection requirements. We hope that the model proposed in the previous section will help them.

Age assurance does not imply the sharing of personal data, such as the date of birth or the age, but the access to the needed information to protect children (access with the meaning of data use, by specific technical, legal or organisational requirements, without necessarily implying the transmission or downloading of data).

In summary, estimation-based approaches often raise specific concerns about 1) inaccuracy caused by errors, 2) data disclosure caused by unnecessary data types or volume (even involving special categories of data), unnecessary processing and unnecessary exposure or 3) unawareness and unintervenability caused by lack of information and control. Verification-based approaches often face challenges related to 1) identification due to the use of identified or identifiable data, 2) inaccuracy caused by issues with document authenticity and user binding or 3) exclusion when documents or certificates are not available.

On the other hand, specific aspects such as the use of unique identifiers, the processing of date of birth or specific age, the need to involve children in the assurance processes, the use of trusted third-party services or the excess of information in the tokens exchanged, to mention just a few examples, make it easier to materialise some of the identified threats. The threat posed by the detection of children through the age assurance solution is always one of the most critical.

Given the eighteen identified threats, some recommendations can be produced for the providers, which are not intended to be a complete or exhaustive set but can serve as an initial element of analysis:

- Linking: Avoid authentication and unique identifiers when possible because it is remarkably effortless to establish links for data subjects who have been authenticated or receive this type of identifier since all requests made within the same session will be connected. Minimise data shared in data flows; each entity should receive only what it needs to do its part of the job. For example, Service requests should not include any information about the content or service provider, and tokens included in Service result data flows should not include more information than passed/not passed (the date of birth or age are not necessary in many use cases).
- Identifying: Avoid processing credentials, attributes, tokens or claims containing identity information; it is not necessary to perform age assurance processes. Anonymous and pseudonymous solutions must be carefully assessed to guarantee robust implementations.
- Inaccuracy: Rely on non-probabilistic methods for age assurance (on age verification methods) whenever necessary, at least in extreme or boundary cases (in doubt). Exhaustively validate the functionality of the processes P1 and P2 and their way of checking age and tokens' content. Eliminate any incentive for fraud, for example, concerning the business model of all the entities involved in age assurance.
- Non-repudiation: Minimise the storage of attributable data and erase this data as soon as possible (remember metadata and logs).
- Exclusion: Provide very different alternatives and advertise them enough. Design easy-to-understand, use and manage solutions. Provide users with the necessary assistance and support.
- Detecting: Design an approach not involving children in age assurance processes; for example, it is the adult who must prove their age to access age-restricted content. Rely on robust encryption so external threat agents may intercept the data flows but not read their content.

– Data Breach: Prepare proactively to fail, to have a data breach. Minimise the collection, processing, and storage of personal data and erase this data as soon as possible (remember metadata and logs). Rely on robust encryption so external threat agents may steal the data but not read their content. Constantly re-assess the proportionate controls using state-of-the-art technologies.
– Data Disclosure and Unawareness: Remember that threats in these two categories must not be able to materialise; they imply regulatory non-compliance. Prevent them from occurring through correct risk management from the point of view of data protection in the organisation: do not collect more data than strictly necessary, ensure that data are not retained longer than strictly necessary, prevent not informed purposes and misuses, provide users with the required information, let them exercise all their rights, do not place them in a situation of urgency, weakness or lack of power in which they must accept any processing, etc.

It is important to note that some threats identified in current solutions cannot be avoided with the available architectures. Therefore, researchers, innovators, and privacy practitioners should explore promising but not yet mature market directions. The first one should be to perform the age assurance process on the user's local devices, making the necessary adaptations to the applications that allow access to content and services or to the operating systems on different devices [6]. The second is to explore the potential of decentralized and user-empowering solutions, such as European digital identity wallets, to solve this problem [12] with the asynchronous approach of the Architecture 3. Furthermore, the third, probably in combination with the previous, is to rely on privacy-enhancing techniques, such as selective disclosure or zero-knowledge proof, to guarantee user unlinkability or anonymity [27].

Concerning regulators and standardization bodies, they should focus on making global and collaborative advances since local efforts in today's digital world will have limited impacts. Standardizing terminology and nomenclature would help in this direction. Regulators should also make clear that the deployment of age assurance solutions must always be aligned with the highest privacy and data protection principles. They should also cooperate with the industry and NGOs (for example, parents or educators associations or professional colleges) to establish global certification schemes, codes of conduct, and regular auditing models which ensure the highest levels of privacy and data protection in the deployed solutions.

Given the potential implications, demonstrated in this research, that all these decisions have, data protection authorities should advise on all these processes, helping to achieve compliance and ensuring enforcement.

## 7   Conclusions

Ensuring children's fundamental rights is paramount, and age assurance solutions play a pivotal role in protecting them from online inappropriate content.

Deploying these solutions following privacy-conscious practices and data protection regulations is crucial. This paper aims to contribute to the ongoing discourse on this subject and foster a safer digital environment for our younger generations without limiting the fundamental rights of all Internet users.

This paper provides a comprehensive analysis of the current age assurance solutions and their potential implications on privacy and data protection. Our innovative threat model, LIINE3DU, identifies specific threats associated with these solutions. The insights gained from our research have significant implications for solution design, policy formulation, and awareness initiatives. Service providers can utilize our threat model to develop more privacy-respecting and compliant age assurance solutions. Policymakers can leverage our findings to shape regulations that promote safety while safeguarding privacy and data protection. Researchers can focus on advancing in the most essential and promising directions. Furthermore, our study provides society with a nuanced understanding of the implications and challenges involved in age assurance, emphasizing the need for privacy-focused solutions.

This paper shows the results of using our threat modelling method for the first time. If new iterations were required because new solutions and architectures must be added to complete the threat model, this might modify the produced threat list. However, the users of this method would work as illustrated in the previous sections, following the same steps in the same way.

We are now working on analysing, in relation to the first categories of threats (LIINEDD, excluding the third D and the U, from a different nature), what risks to the rights and freedoms of users each threat specifically entails and how the probabilities and impacts associated with these risks could be mitigated. This would be essential to adopt a risk-based approach, so common in the data protection field to decide about the implementation of the appropriate technical and organisational measures to integrate the necessary safeguards into the processing in order to meet the requirements of regulation, to ensure a proper level of security, etc.

**Acknowledgement.** This study was funded by the Agencia Española de Protección de Datos (AEPD), the Spanish Data Protection Authority. The authors want to thank the entire AEPD team and specifically the staff of the Innovation and Technology Division for their involvement in this project and all their generous comments and contributions.

**Disclosure of Interests.** The authors declare that they have no known competing financial interests or personal relationships that could have appeared to influence the work reported in this paper.

# References

1. Directive 2018/1808 of the European Parliament and of the Council of 14 November 2018 amending Directive 2010/13/EU on the coordination of certain provisions laid down by law, regulation or administrative action in member states concerning the provision of audiovisual media services (Audiovisual Media Services Directive) in view of changing market realities (2018). https://eur-lex.europa.eu/eli/dir/2018/1808/oj
2. Age Appropriate Design Code (2022). https://californiaaadc.com/
3. Regulation 2022/2065 of the European Parliament and of the Council of 19 October 2022 on a single market for digital services and amending Directive 2000/31/EC (Digital Services Act) (2022). https://eur-lex.europa.eu/eli/reg/2022/2065/oj
4. Online Safety Act (2023). https://www.legislation.gov.uk/ukpga/2023/50/enacted
5. 5RightsFoundation. But how do they know it is a child? Age assurance in the digital world (2021). https://5rightsfoundation.com/uploads/But_How_Do_They_Know_It_is_a_Child.pdf
6. AEPD. Protection of minors on the Internet- Technical note with the description of the proofs of concept (2023). https://www.aepd.es/guides/technical-note-proof-of-concept-age-verification-systems.pdf
7. Bertrand, A., Diaz, M.C., Hair, E.C., Schillo, B.A.: Easy access: identification verification and shipping methods used by online vape shops. Tobacco Control (2024)
8. Blake, P.: Age verification for online porn: more harm than good? Porn Stud. **6**(2), 228–237 (2019)
9. Brennen, S., Perault, M.: Keeping kids safe online: how should policymakers approach age verification? The Center for Growth and Opportunity (2023)
10. BSI. PAS 1296:2018 - Online age checking. Provision and use of online age check services. Code of Practice (2018). https://knowledge.bsigroup.com/products/online-age-checking-provision-and-use-of-online-age-check-services-code-of-practice?version=standard
11. CNIL. Online age verification: balancing privacy and the protection of minors (2022). https://www.cnil.fr/en/online-age-verification-balancing-privacy-and-protection-minors
12. EC. Second meeting of the Task Force on age verification (2024). https://digital-strategy.ec.europa.eu/en/news/second-meeting-task-force-age-verification
13. EDRi. Online age verification and children's rights (2023). https://edri.org/our-work/policy-paper-age-verification-cant-childproof-the-internet/
14. Egan, K.L., Villani, S., Soule, E.K.: Absence of age verification for online purchases of cannabidiol and delta-8: implications for youth access. J. Adolesc. Health **73**(1), 195–197 (2023)
15. euCONSENT. euCONSENT project (2024). https://euconsent.eu/home-euconsent-project/
16. de Farias, J.C.L.A., Carniel, A., de Melo Bezerra, J., Hirata, C.M.: Approach based on STPA extended with STRIDE and LINDDUN, and blockchain to develop a mission-critical e-voting system. J. Inf. Secur. Appl. **81**, 103715 (2024)
17. Georgiadis, G., Poels, G.: Towards a privacy impact assessment methodology to support the requirements of the general data protection regulation in a big data analytics context: a systematic literature review. Comput. Law Secur. Rev. **44**, 105640 (2022)

18. ICO. Age appropriate design: a code of practice for online services (2024). https://ico.org.uk/for-organisations/uk-gdpr-guidance-and-resources/childrens-information/childrens-code-guidance-and-resources/age-appropriate-design-a-code-of-practice-for-online-services/
19. ISO. ISO/IEC WD 27566-1 Information security, cybersecurity and privacy protection -Age assurance systems- Framework (2023). https://www.iso.org/standard/88143.html
20. Jarvie, C., Renaud, K.: Are you over 18? A snapshot of current age verification mechanisms. In: Dewald Roode Workshop (2021)
21. LINDDUN. Privacy threat modeling (2024). https://linddun.org/
22. Nash, V., O'Connell, R., Zevenbergen, B., Mishkin, A.: Effective age verification techniques: lessons to be learnt from the online gambling industry. Available at SSRN 2658038 (2012)
23. Nweke, L.O., Abomhara, M., Yayilgan, S.Y., Comparin, D., Heurtier, O., Bunney, C.: A LINDDUN-based privacy threat modelling for national identification systems. In: Proceedings of the IEEE Nigeria 4th International Conference on Disruptive Technologies for Sustainable Development, pp. 1–8 (2022)
24. OECD. Children in the digital environment: revised topology of risks (2021). https://doi.org/10.1787/9b8f222e-en
25. Pasquale, L., Zippo, P., Curley, C., O'Neill, B., Mongiello, M.: Digital age of consent and age verification: can they protect children? IEEE Softw. **39**(3), 50–57 (2020)
26. Robles-González, A., Parra-Arnau, J., Forné, J.: A LINDDUN-based framework for privacy threat analysis on identification and authentication processes. Comput. Secur. **94**, 101755 (2020)
27. Ronis, J.: Don't trust when you can verify: a primer on zero-knowledge proofs (2024). https://www.wilsoncenter.org/article/dont-trust-when-you-can-verify-primer-zero-knowledge-proofs
28. Sas, M., Mühlberg, J.T.: A risk-based evaluation of available and upcoming age assurance technologies from a fundamental rights perspective (2024). https://www.greens-efa.eu/en/article/study/trustworthy-age-assurance
29. Williams, R.S., Phillips-Weiner, K.J., Vincus, A.A.: Age verification and online sales of little cigars and cigarillos to minors. Tobacco Regulat. Sci. **6**(2), 152 (2020)
30. Wuyts, K., Sion, L., Joosen, W.: LINDDUN go: a lightweight approach to privacy threat modeling. In: Proceedings of the IEEE European Symposium on Security and Privacy, pp. 302–309 (2020)
31. Xiong, W., Lagerström, R.: Threat modeling-a systematic literature review. Comput. Secur. **84**, 53–69 (2019)
32. Yar, M.: Protecting children from internet pornography? A critical assessment of statutory age verification and its enforcement in the UK. Policing: Int. J. **43**(1), 183–197 (2020)

# Access Your Data... if You Can: An Analysis of Dark Patterns Against the Right of Access on Popular Websites

Alexander Löbel[1]($\boxtimes$), René Schäfer[1], Hanna Püschel[2], Esra Güney[1], and Ulrike Meyer[1]

[1] RWTH Aachen University, Aachen, Germany
`{loebel,meyer}@itsec.rwth-aachen.de, rschaefer@cs.rwth-aachen.de,`
`esra.gueney@rwth-aachen.de`
[2] TU Dortmund University, Dortmund, Germany
`hanna.pueschel@tu-dortmund.de`

**Abstract.** Various regulations including the GDPR empower users with the right to request a copy of their personal data processed by data holders. This *right of access* can serve as the foundation of exercising other data subject rights, including erasure and rectification of the processed data. Like other regulations, the GDPR does not prescribe any specific procedure data holders need to implement to handle data subject access requests but requires them not to erect any material or formal hurdles in the assertions of their rights. In this paper, we focus on popular online service providers as data holders and investigate in which form they allow users to make data access requests directly on their websites and whether they use any strategies to impede such requests. Our systematical analysis of the process of submitting access requests on 166 account-based websites from the top 500 entries of the Tranco list reveals 238 instances of dark patterns impeding the submission of data subject access requests on 113 (68%) of the examined websites.

**Keywords:** right of access · dark patterns · usable privacy

## 1 Introduction

Data protection regulations around the world currently govern the collection and processing of personal data, including the *California Consumer Protection Act* (CCPA) [14] or the *General Data Protection Regulation* (GDPR) [21]. These regulations aim to enhance data transparency and empower users with rights to control the use of their personal data. These rights include the right to access, the right to erase, and the right to rectify personal data. While each of these rights is granted independently, exercising the right of access may often be the initial step to exercise other rights. The GDPR does not require data holders to follow any specific technical procedure to grant these rights, but rather only generally *Art. 12 para. 2 GDPR* requires them to facilitate the exercise of data subject rights without material or formal hurdles.

M. Jensen et al. (Eds.): APF 2024, LNCS 14831, pp. 23–47, 2024.
https://doi.org/10.1007/978-3-031-68024-3_2

Previous studies have emphasized the importance of the right of access [43,47,48], yet challenges persist in its implementation. Studies involved users requesting data from controllers [2,10,46,57,58], researchers initiating access requests [11,39,54,59,61,67,68], and examinations of the usability of returned data [70]. These studies consistently highlight issues with data subject access request (DSAR) submissions and controller compliance. Notably, Pöhn et al. [59] described dark patterns hindering DSARs with 27 data controllers. Such dark patterns – deceptive design choices tricking users into behaving differently than originally intended – have been documented across various domains [5,16,18,24, 25], such as in shopping platforms [50], social networking sites [52], games [73], and mobile platforms [27,32]. Particularly concerning privacy and transparency, dark patterns have been observed extensively [8,30,34,36,40]. Consent banners stand out as a popular domain for dark pattern usage [7,29,38,45,55,66,69], prompting researchers to automate checks and countermeasures for consent banner-related dark patterns [4,28,35].

Existing research has not focused on a particular way to submit a DSAR. Nonetheless, online services have to inform users about data subject rights. However, there is no detailed analysis of whether service providers leverage the control they have over their own websites to hinder DSARs. We concentrate on online service providers and on how they allow users to submit DSARs on their websites. Specifically, we explore whether popular online service providers utilize dark patterns that hinder DSARs on the website itself. Through this, we contribute to the examination of dark patterns in transparency by expanding upon the data of Pöhn et al. [59] through a larger-scale study. Additionally, we utilize the EDPB taxonomy [19] to categorize our findings, facilitating the mapping of identified dark patterns to GDPR articles potentially breached. Overall, we explore the following research questions:

1. What mechanisms are available for users to submit a DSAR on the websites of popular online service providers?
2. Do these providers employ dark patterns that hinder the submission of DSARs on their websites?
3. Which types of dark patterns, based on the EDPB taxonomy, are most common on popular websites while posing DSARs?

To answer these questions, we systematically analyzed the process of finding a method to submit a DSAR directly on 166 account-based websites from the top 500 entries of the Tranco list [42]. Our analysis reveals 238 instances of dark patterns on popular websites. We provide a fully labeled dataset with descriptions, categorizations into the EDPB taxonomy, and screencasts of our request analysis for each website to enhance transparency of our research and facilitate further research in this area.[1] Note that we deliberately decided not to answer request confirmation e-mails in order to keep manual work for website operators low and we do not examine the returned data but rather the existence of dark patterns until DSAR submission.

---

[1] https://osf.io/4jvhx/.

# 2    Background and Related Work

We discuss regulations governing personal data usage, with a focus on the GDPR as we conduct the study from within the EU. We detail prior work investigating the right of access, introduce our working notion of dark patterns, outline relevant works about dark patterns, and discuss pertinent taxonomies.

## 2.1    The Right of Access

The right of access was included in data protection regulations around the world as early as the 1960s [12] and became a cornerstone of data protection law, which it has retained to this day [48]. The right of access was also established for the member states of the EU in *Art. 15* of the GDPR. The GDPR initially only referred to the EU but was adopted for the EEA [15]. The GDPR is often seen as a substantial influence for other data protection laws around the world, e.g., in the US, Brazil, Japan, and Switzerland, which also provide a right of access [26,31]. Although the GDPR is a regional data protection law and does not apply worldwide, its scope of application is enormous [47]. The GDPR regulates the territorial scope of application itself in *Art. 3*. Accordingly, the GDPR applies to data processing activities within the EU and includes data subjects located in the Union, even if processing occurs outside the Union (*Art. 3 para. 1 GDPR*) and even data processing by controllers not established in the EU is covered by the GDPR under certain conditions (*Art. 3 para. 2 GDPR*). The right of access granted by *Art. 15* provides data subjects an insight into whether and how their data is being processed. The reasoning for this is that fair and transparent processing is only possible if the data subject can obtain information about the processing (*Recital 60*). Other data subject rights can be aggravated if the data subject does not know what data the controller is processing. In addition, the controller shall also provide a copy of the personal data undergoing processing in accordance with *Art. 15 para. 3* of the GDPR.

**Data Subject Access Requests.** This right must be asserted by submitting a corresponding request to the controller. However, *Art. 15* of the GDPR does not describe how the right to information should be asserted. The request can, therefore, be made without any formal requirements. The process of making a request is at most specified by *Art. 12 para. 2 GDPR*, which stipulates that the controller must facilitate the exercise of the data subject's rights in accordance with *Art. 15*. Additionally, *Recital 59* states that mechanisms should be established to ensure data subjects can request and obtain access to personal data free of charge. These legal requirements are not particularly specific, as the GDPR deliberately aims to be technology-neutral. In any case, no material or formal hurdles may be erected in the assertion of data subjects' rights. A violation is therefore assumed if access to the information is made considerably more difficult without objective reasons, or if the information can only be obtained by accepting a media breach.

**Studies on the Right of Access.** In practice, the process of submitting DSARs seems to be hampered. One problem constitutes user verification, which should ensure that a user is authorized to the collected data. There are works [11,13,17,49,56] showing one can abuse the right of access to learn personal information of others. Lauradoux [41] shows that this is facilitated for a governmental adversary. In [6,65], data controllers' perspectives are explored, and recommendations for authorization of data subjects are examined. However, these recommendations can lead to doubtful authentication mechanisms [6]. Multiple works show that the implemented processes to submit a DSAR lack usability. Despite the right of access forming the basis for subsequent rights, it has been shown that it falls short of expectations [43]. This is further evidenced by empirical studies investigating the process of submitting DSARs by researchers [54,61,67,68], or with the help of volunteers [2,10,46,57] and students [3].

Not only the returned data is often impractical and non-structured for users struggling to make sense of the data [70] but also the path to submit a request seems to be non-ideal [58], clustered with obstacles and data controllers that do not respond adequately [10,39,59,68]. Waldman [71] raises the assumption that such inconveniences may be intentional and things like privacy dashboards or consent choices are merely symbolic and primarily maintain the appearance that the industry cares about privacy. Similarly, we presume that some online services try to implement dark patterns hindering users from accessing their data, where these online services have control, i.e., on their websites. We still remark that it is often not possible to reach a reliable decision of whether a pattern was actually implemented to hinder users, was a sincere programming error, or serves another purpose justifying such obstructions.

Nonetheless, Pöhn et al. [59] describe occurrences of dark patterns they found in their exploration of DSAR submissions. Our work complements their work by specifically looking into dark patterns hindering a request submission on the websites of popular services on a larger data set. Furthermore, we categorize our findings into a pertinent taxonomy, providing references to the GDPR articles the found dark patterns might breach. We confirm some of the patterns described by Pöhn et al., e.g., *making it impossible to access GDPR requests* [59] often translates to the dark pattern *Dead End* of the EDPB taxonomy [19] that we found multiple times. Before we describe our methodology, we introduce dark patterns and where they can be found in the context of transparency.

## 2.2   Dark Patterns

Dark patterns[2] are malicious user interface design strategies that influence the decision-making of users in favor of an online service [25,50]. The term *dark pattern* was coined by Brignull in 2010, who created a website[3] showcasing actively used dark patterns. Deceptive designs have been shown in various contexts on

---

[2] Sometimes also referred to as *deceptive patterns* or *deceptive designs*.
[3] https://www.deceptive.design/ *(accessed 31.01.2024)*.

the web [16,18,27,32,36]. In addition to research papers, governmental reports and guidelines on dark patterns exist. In their report[4], the *Norwegian Consumer Council* explains how companies use dark patterns to nudge people into choosing privacy-intrusive user settings. Additionally, an extensive report published by the European Commission [18] analyzed dark patterns and stressed their prevalence on websites and in apps. This ubiquity of manipulative designs led to diverse countermeasures, including user awareness [5,16,53], automatic detection [4,35,50,66], and visual countermeasures [64]. Bongard-Blanchy et al. [5] categorized countermeasures by splitting them into four regions of action (*educational, design, technical,* and *regulatory*) and four intervention scopes (*awareness, detection, resisting,* and *elimination*). However, as technical countermeasures are an arms race between developers and service providers, researchers and experts still underline the need for stronger regulations [33,63,72].

**Legislation on Dark Pattern Usage.** One response to such calls is the *Digital Services Act* (DSA) [22]. It explicitly includes a ban on dark patterns. *Art. 25* of the DSA prohibits arrangements whereby users are deceived in their decision-making. Although the DSA already came into force in November 2022, most of the provisions, including *Art. 25*, only apply from February 2024 (*Art. 93*). We conducted the study before February 2024, and hence service providers did not have to comply with these provisions yet. Thus, we cannot say how this new regulation will affects our findings in the future. There are more general works on the legality of dark patterns as well. The website of Brignull provides an overview of passed laws concerned with dark patterns and manipulative designs. Additionally, Luguri and Strahilevitz [44] argue that some dark patterns can be considered unlawful.

**Dark Patterns Undermining Transparency.** There are works (e.g. [7,37,38, 45,66]) that provide evidence supporting the claim about the industry's unwillingness by looking at the design of consent choice banners. Note that cookie banners are covered primarily by the ePrivacy Directive *Art. 5 para. 3* [20]. Krisam et al. [38] reviewed consent choice banners on Germany's top 500 visited websites and found that 85% used visual nudges to make people accept cookies. Nouwens et al. [55] scraped consent pop-ups on 680 websites in the UK and found that only 12% met basic requirements regarding European law. The authors additionally confirmed the influence of designs of commonly used consent choice banners in a user study with 40 participants. This agrees with Habib et al. [29], who investigated cookie consent forms in an online user study and derived design implications regarding usability. Further research includes the automatic detection of dark patterns in consent notices [28,35] and user behavior dealing with dark patterns in cookie banners [5,69]. Apart from consent choice banners, dark patterns were also found in other parts of websites,

---

[4] https://storage02.forbrukerradet.no/media/2018/06/2018-06-27-deceived-by-design-final.pdf *(accessed: 20.01.2024).*

such as account management [34], pages concerned with opting-out of advertisement options [30], or pages dealing with legitimate interest [40]. Bösch et al. [8] created a general framework of *privacy dark patterns* and showed how they follow common templates.

**Dark Pattern Taxonomies.** As an ever-growing number of dark patterns is identified, several taxonomies [51] try to provide an overview of the situation by grouping dark patterns by their characteristics. These taxonomies focus on, inter alia, games [73], shopping [50], social network services [52], attention-capture dark patterns [53], and privacy [8]. One well-established taxonomy is given by Gray et al. [25] and was extended and adapted to the context of shopping by Mathur et al. [50]. Additionally, researchers collected definitions and types of dark patterns from existing taxonomies and included further research and governmental reports [44,51], leading to more general taxonomies. A recent work by Gray et al. [24] combined this knowledge and clustered dark patterns into high-level, meso-level, and low-level.

While the taxonomies and reports above help to understand the range of dark patterns in various contexts, they are less helpful for scenarios closely related to the right of access. Here, the taxonomy of the EDPB [19] is applicable when investigating account-based websites that are subject to rules of the GDPR. The taxonomy maps manipulative website parts to concrete dark patterns sorted into six categories *Overloading*, *Skipping*, *Stirring*, *Obstructing*, *Fickle*, and *Left in the Dark*. Each category contains 2–4 dark patterns, totaling 16 dark patterns. A detailed description follows in Sect. 3.3. Compared to [8], it offers a more nuanced differentiation between individual dark patterns regarding the GDPR and is well received within the dark pattern community [1,9]. Overall, the taxonomy by the EDPB is suitable for our study due to its proximity to the GDPR.

Summarizing, dark patterns seem to be a popular tool to subvert user's choices online. A body of work has documented the real-world use of deceptive designs, ranging from shopping to transparency and privacy. In the realm of privacy and transparency, the prevalence of dark patterns, especially in consent choice banners, has been well-researched. We add to this by investigating the process of submitting a DSAR on popular websites for dark patterns. While Pöhn et al. [59] also described the dark patterns they encountered while investigating DSAR submissions, we complement this data with a larger-scale analysis and a categorization of dark patterns into a fitting taxonomy providing references to the GDPR articles that might be breached.

## 3   Methodology

Following, we outline the methodology for examining the prevalence of dark patterns that can hinder a user's ability to request a copy of their personal data directly on a visited popular website. We first describe the process of creating our dataset of popular websites and the exclusion criteria used. Then, we detail our systematic process for the analysis on each website.

## 3.1   Website Corpus

The most clear ways to argue that a controller has to adhere to the GDPR are as follows. Either the establishment of the controller itself is located in the EU (*Art. 3 I GDPR*) or the service is offered to users in the EU (*Art. 3 para. 2 lit. a) GDPR*). This constitutes one of the reasons why we exclude any website without native English interface (mostly excluding websites with main user base in Asia or Russia). Since most service providers offer a native English interface, limiting our research to said services allows us to compare our results, especially regarding dark patterns that are not purely visual. Another reason for exclusion was the service being non-reachable since we access the websites from a country within the EU. Furthermore, consider that the GDPR also applies if citizens from the EU are tracked on the service according to *Art. 3 para. 2 lit. b)*. As we consider the top 500 entries of the Tranco list - hence highly popular online services - we argue it is highly probable that the service offered is directed to EU citizens (too), as long as they also have a native English interface. While this does not always imply that a website has to adhere to the GDPR, the impact of the GPDR on regulations in other regions of the world analyzed in [47] supports the assumption that these websites fall under similar regulations. Hence, we do not include additional checks that websites must adhere to a fitting regional data policy.

Furthermore, we excluded websites requiring a verified phone number to register to protect the researchers' privacy[5]. For other personal information, such as addresses, we use fictitious data. Websites without a way of creating an account were also excluded. Those were usually governmental or educational websites. Lastly, we excluded websites using an account basis already present in the set of included websites. Those were usually services provided by bigger companies such as Google or Microsoft. Concluding, we excluded websites meeting any of the following criteria:

- The website was not reachable.
- The website lacked a native English user interface.
- No method for creating an account was found.
- A verified phone number was required for account creation.
- The website obviously utilized the account basis of another service already included in our list.

We started with a list of the 500 most visited websites according to the Tranco list[6] [42] generated on 18.07.2023. Three researchers independently accessed each website from the same country within the EU and following the exclusion criteria resulted in 166 websites to be considered. The complete list of 500 websites, including the exclusion reasons, can be found in the OSF repository[7]. The distribution of the 166 websites according to the "Website Categorization API" of

---

[5] This was only the case for five websites. Hence, we did not obtain an anonymous phone number to create an account.

[6] https://tranco-list.eu/list/W9Z49 *(accessed: 18.07.2023)*.

[7] https://osf.io/4jvhx/.

WhoisXMLAPI[8] is shown in Fig. 1. Most websites belong to the category "Computer and Internet Info" (35), closely followed by "Business and Economy" (26) and "News and Media" (19). Despite this skew, we have 33 different categories in our dataset, providing some diversity.

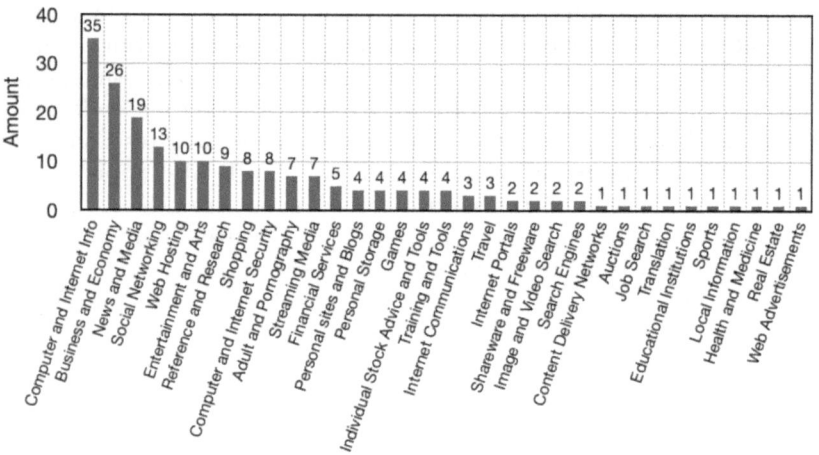

**Fig. 1.** Distribution of included websites per category. In total, there are 33 different website categories present in the dataset.

## 3.2 Request Procedure

For each website, we analyzed the path to pose a DSAR within the desktop version of the website. Each of the three researchers created a new account with an e-mail address created only for this study and fictitious data (names, addresses,... ). Each researcher carried out the request as far as possible on the website to not miss any interaction on the website itself. However, we did not respond to confirmation e-mails to avoid creating unnecessary work for data protection officers. By this, the amount of work created by our study is negligible, since most websites that do not automate the DSAR response require at least confirmation through the registered e-mail address before responding. In the case of download or request buttons (explanations for the request mechanisms encountered follow in Sect. 3.3), we clicked the button as we otherwise could never be sure whether there is further action needed on the website and thus potentially miss additional hurdles. Note that these also often require confirmation via e-mail or lead to automated responses.

The accounts used did not generate a lot of data through usage of the service since they were often created minutes ago. However, we do not want to investigate the returned data but rather only the existence of dark patterns

---

[8] https://whois.whoisxmlapi.com/ (accessed: 13.02.2024).

*until* the DSAR submission. Hence, our analysis is not hindered by this. For account creation and request analysis, each researcher accessed the website from the same country within the EU again. All researchers independently performed their request analysis. Data collection took place from July 2023 to August 2023. To achieve comparability between researchers for the later discussion, a search protocol was created and followed to navigate each website. While we suspect many users would first use a search engine to find out how they can request their data, we investigate the online infrastructure of the services themselves. Hence, we do not start with arbitrary vantage points on which a user could land after using a search engine, but rather with the index page. This is to ensure we find dark patterns regardless of the vantage point, but do not claim a user would necessarily encounter every single dark pattern we found. To achieve such a search protocol, two researchers discussed how they imagined a user searching for the option to request their data on the website. Then, both researchers picked ten random websites from the list and tried to find a way to pose a DSAR through the website using the initially discussed search protocol. The protocol was re-evaluated and updated accordingly. The final protocol can be seen in Fig. 2 and consists of the following steps.

1. Log into the website and check whether an account dropdown menu exists. If so, match the entries to the predefined set of keywords from top to bottom. If we find a matching entry, click on it and repeat the search there. Once we cannot find any matching keywords anymore, or if it is obvious that we cannot find anything helpful here, stop the search.
2. If a settings menu entry exists, click on it and start the keyword search again.
3. If a dedicated profile settings menu entry exists, click on it and start the keyword search again.
4. If we did not find anything helpful until now, we click through all settings menu entries in a breadth-first search manner, i.e. we skim through each of them to look for related controls placed in unexpected menu points.
5. As a last resort, we check the website's footer and match for keywords again.

After finalizing the search protocol, we used it as guidance for all subsequent request attempts. However, note that it is not feasible to always exactly adhere to this protocol for a corpus of 166 websites because of differences in the interfaces of the websites and derivations introduced by people navigating the web in slightly different ways. This protocol is intended to allow for general comparability between the attempts of the three researchers for the later analysis we describe in Sect. 3.4. Additionally, we made screencasts of each request attempt per researcher, which we used in the discussion to enrich the researcher's notes and allow us to make our research more transparent. The screencasts are also available at the OSF repository.

## 3.3 Codebook

In the following, we explain the codebook we used to analyze and categorize the request mechanisms and the dark patterns.

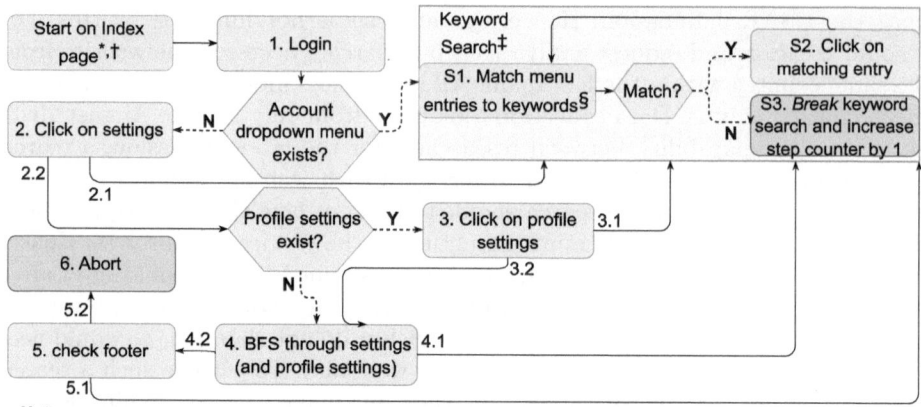

**Notes**
*: On each page load, we skim the page for obvious ways to carry out an access request. If there is an obvious way, it preceedes any other step.
†: If it is necessary to fill out a form, we fill it as far as possible.
‡: When we match for keywords, we do it always in a depth-first and top-to-bottom manner.
§: <u>Keywords</u>: access request, download request, data rights, personal data, privacy rights, GDPR, exercise your rights, privacy, security, compliance

**Fig. 2.** The final protocol we used to guide the search for ways to pose a DSAR.

**Request Mechanisms.** Because of slight derivations in website usage, different researchers could find different mechanisms. For each website, we thus classified the final results of the request into the following mechanisms:

- *Simple Form*: any form that can be submitted by just using click events
- *Form*: any form that needs more input than just click events to submit it
- *E-Mail*: stating an e-mail address to which one is supposed to write to make a request
- *Request Button*: button to start a data access request without further input
- *Download Button*: button to instantly download your data
- *Form Requiring Personal Data*: forms that require data that we did not want to disclose (e.g., a verified telephone number)
- *No Information Found*: no information how to pose a request was found
- *Impossible To Complete*: could not go on with the information provided
- *Not Checked*: the website was not checked

**Dark Patterns.** As we are not only interested in the DSAR submission mechanisms that users are provided on these popular websites, but specifically in the dark patterns implemented, we touch upon the categories of the EDPB taxonomy [19] in the following and introduce the types of dark patterns present in each category. Concrete examples are included in Sect. 4.4.

Overloading: Dark patterns within this category confuse users by providing *Too Many Options* regarding privacy choices, creating a *Privacy Maze* where users are led in circles, or using *Continuous Prompting* to make users enter more personal data than they initially intended.

Skipping: Dark patterns that use skipping try to distract users by using pointers to different page elements (*Look Over There*), making them overlook privacy choices by defaulting to unwanted options (*Deceptive Snugness*).

Stirring: Stirring is used for *Emotional Steering* where visual information or text influences users' emotions. This includes making buttons look deactivated or mentioning that posing a privacy request might lead to additional costs. Additionally, stirring can be accomplished by placing information or controls *Hidden in Plain Sight*. For example, by burying an unstyled contact e-mail address in a large paragraph of text.

Obstructing: These patterns are rather aggressive as they actively hinder users in their process to exercise their rights by making the required steps or the path *Longer Than Necessary*, or by using *Misleading Actions* where a discrepancy of available information and performable actions confuse users. For example, a site states that users can exercise their rights by clicking on a given link, which takes them to a privacy article without any options to submit a request as promised. A *Dead End* can prevent users from continuing with their request. This includes misfunctioning buttons and websites where participants get stuck without further information on how to continue.

Fickle: Dark Patterns within this category affect the structural integrity of the website itself, through *Lacking Hierarchy* or having an *Inconsistent Interface*, also including *Language Discontinuity* where parts of a site suddenly switch the language without identifiable reasons. Additionally, *Decontextualising* works by hiding controls in unrelated sections or tabs and is used to make users overlook actionable steps.

Left In The Dark: As the name suggests, dark patterns in this category confuse users by providing *Ambiguous Wording or Information* or *Conflicting Information*, where two pieces of information contradict each other. Through this, users usually do not know how to continue or whether they are following a sensible path to submit a request.

### 3.4   Analyzing Request Attempts

To analyze the runs for each website per researcher, we followed a qualitative analysis approach [23]. Starting with an open coding approach, each researcher independently noted the timestamps of the screencasts where they felt they saw a deceptive behavior by the website and added a comment describing the manipulation. Afterward, all three researchers sat together to coalesce the independent analyses into a final codebook. For each website and noted possible manipulation, it was discussed whether it

1. actually constitutes a dark pattern according to our notion,
2. in which category of the codebook it belongs,
3. and which specific type of dark pattern it is.

The discussion continued until consensus. If necessary, the respective screen-cast was reviewed together again. In the independent coding phase, the taxon-omy's exact dark pattern was not yet stipulated. Rather, notes were created as basis for the discussion. Note that because of this, it is not possible to calcu-late inter-rater agreement scores on the results of the independent coding phase. The discussion was held over multiple meetings in September 2023. In the end, we reached a fully coded table of dark patterns per website, together with a timestamp where one can see the encountered dark patterns in the recordings, a description for each dark pattern, and a categorization as discussed by the three researchers. We further discuss our findings in the following chapter based on this codebook.

## 4   Results

We first present the request mechanisms identified (RQ1). Then, we focus on the prevalence of dark patterns across the investigated websites (RQ2). Subse-quently, we present the time spans required to finish all actions on a website for a DSAR to check the impact of dark patterns on request difficulty. Finally, we outline the distribution of dark patterns (RQ3) based on the EDPB taxonomy.

### 4.1   Request Mechanisms

Figure 3 illustrates the 265 identified request mechanisms. Note that it was not always the case that all researchers arrived at the same outcome. On 33 (19.9%) websites, we found more than one mechanism to pose a DSAR. Each share of the left pie chart shows the share of one of the mechanisms explained in Sect. 3.3. Most prevalent were online forms (26%), closely followed by an e-mail address (24%) to which one should write to submit a request. The smaller portions include simple forms (9%), request buttons (8%), and direct download buttons (3%). In 15% of the examined websites, at least one of the researchers could not find sufficient information. The right chart in Fig. 3 shows on how many websites either one, two, or all three researchers could not find sufficient information.

### 4.2   Dark Patterns per Website

Of the 166 examined websites, 113 (68%) contained at least one dark pattern, totaling 238 instances. On average, a website contained 1.43 dark patterns with a standard deviation of 1.59. Figure 4 illustrates the number of websites against the number of dark patterns found on a website. We see a trend, with many websites having a few patterns, gradually decreasing to a few extreme cases. Approximately one-third (32%) of the examined websites showed no dark pat-terns.

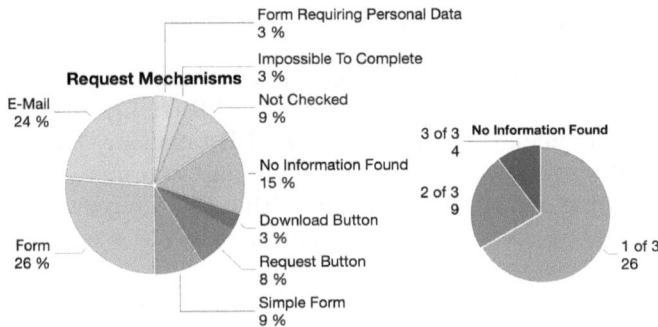

**Fig. 3.** Distribution of the 265 request mechanisms found on the 166 examined websites together with the amount of websites where one, two, or all three researchers did not find sufficient information for a DSAR.

## 4.3   Timing Measurements

Figure 4 also depicts box plots for each class containing websites with a specific number of dark patterns. The box plots represent the time a researcher needed to finish the actions on a website to pose a DSAR. Each attempt by any researcher is included as data point, resulting in a maximum number of three timing measurements per website. We can see a trend of the median and, in most cases, the upper and lower percentiles increasing with the number of dark patterns. However, box plots with fewer dark patterns have more outliers, possibly due to more websites with fewer dark patterns (as shown in Fig. 4) leading to more timing measurements in that plot.

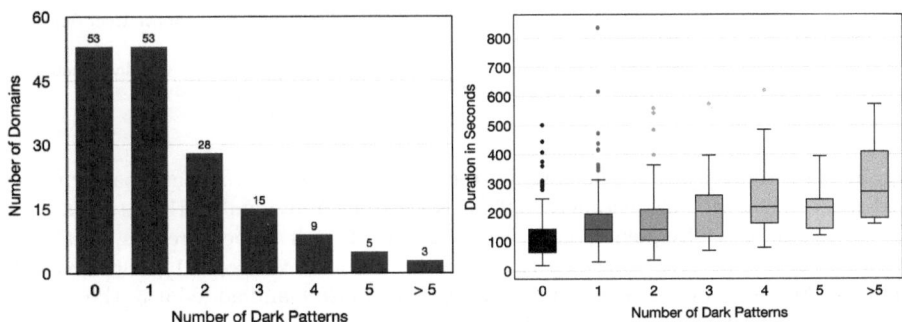

**Fig. 4.** The usage of dark patterns from all websites within our study (left) and timing measurements for each of the website classes (right). Overall, we found that about 68% of all investigated websites utilize at least one dark pattern. Additionally, we observe a trend of taking more time for a DSAR the more dark patterns are present on a website.

## 4.4   Dark Pattern Distribution

As previously noted, we identified 238 instances of dark patterns in our examination of 166 popular websites. Figure 5 displays the distribution of these dark patterns across the six main categories of the EDPB taxonomy [19] (see Sect. 3.3). Below the main categories, a pie chart for each main category shows the number of occurrences of the named dark patterns within the category. In the following, we show examples of dark patterns documented in our study for each of the main categories.

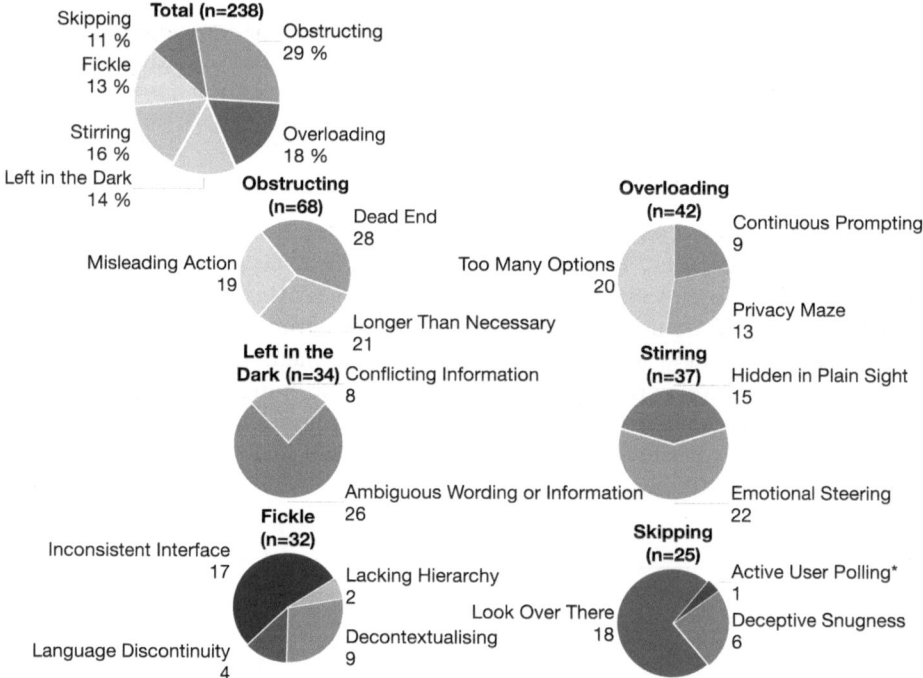

**Fig. 5.** Distribution of all 238 dark patterns found. *Obstructing* dark patterns were used most often with 68 instances (29%). Regarding individual dark patterns across categories, the most frequently used were *Dead End* with 28 occurrences, *Ambiguous Wording* (26), *Emotional Steering* (22), *Longer Than Necessary* (21), and *Too Many Options* (20). We found a new pattern *Active User Polling* and added it to the *Skipping* category in the EDPB taxonomy [19].

**Obstructing.** The most prevalent category is *Obstructing*, which is nearly evenly split into the three available dark patterns. With 28 cases, *Dead End* is the most common. Examples include services limiting the number of requests a user can submit in a long period, while providing a shorter timeframe for downloading the data. One example is `epicgames.com`, which gives the user three days to download the data once it is prepared. Simultaneously, the user can request it only once every 90 days. Note that policies such as the GDPR often allow for limiting the frequency of requests to prevent services from being overburdened by a mass of requests from a single user. Nonetheless, the inability to retrieve their data if the user missed the shorter timeframe until the time for another request to be allowed makes this combination a *Dead End*. In one case (`soundcloud.com`), a request form could not be completed because it claimed the provided e-mail address was invalid, despite being the exact e-mail address used for registration. Another variant of this pattern involves broken links leading nowhere, as defined by the EDPB taxonomy [19]. A remarkable case of unclickable links were links shown on an image within a privacy policy on `quora.com`, allowing users to click only the image, but not the included links. Some websites lead to a *Dead End* by requiring identifiers not all users necessarily possess (such as booking numbers on `booking.com`) as a means of identification, limiting DSARs to only a subset of users (such as paying customers).

Examples of *Misleading Actions* include websites such as `salesforce.com` stating in their privacy policy that users can exercise their rights in the account settings, while no controls are available there. In other instances, we were misled when attempting to change the language of the privacy policy back to English because it defaulted to another language, resulting in redirection to the start page. Examples of *Longer Than Necessary* (22 cases) involve service providers forcing users to make DSARs for each product individually instead of providing the option to request all stored data at once. On the website `xiaomi.com`, you have to wait for a 60-second timer before being able to click on the request confirm button (Fig. 6a). Another example is the buttons on `tumblr.com` that look like they are disabled (Fig. 6b).

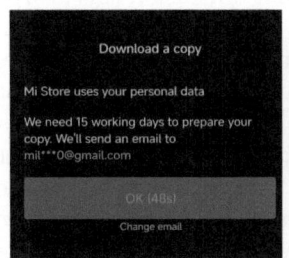
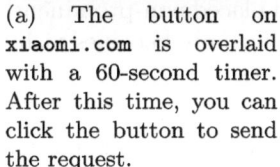
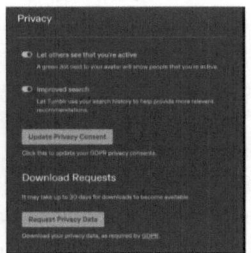
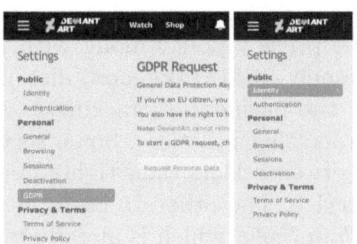

(a) The button on `xiaomi.com` is overlaid with a 60-second timer. After this time, you can click the button to send the request.

(b) The two buttons on `tumblr.com` look greyed out, suggesting they are disabled. However, they are working.

(c) The menu item "GDPR" on `deviantart.com` is only visible on a specific URL found in the privacy policy. Clicking on any other setting hides the menu item again.

**Fig. 6.** Examples of dark patterns found.

**Overloading.** The second-largest category is *Overloading* (18%), with nearly half of the dark patterns as *Too Many Options*. This often occurred on websites providing a plethora of links or documents seemingly related to privacy. In one extreme case, on `cisco.com`, we found a link directing to a database of "Privacy Sheets" containing documents in multiple languages for any product they offered. This is followed by the closely related *Privacy Maze* with 13 cases, often resulting from many different links linking back and forth, resulting in users going in circles. *Continuous Prompting* is primarily executed through DSAR forms asking for unnecessary personal information such as telephone numbers, which might discourage privacy-aware users from submitting.

**Left in the Dark.** The category *Left in the Dark* has a share of 14% of the total number of dark patterns. The main dark pattern encountered here was presenting users with *Ambiguous Wording or Information*. For example, some services offer a way to "export their data" while hinting the returned data may be incomplete, e.g., on `bbc.com` where one is told "they won't include everything" without specifying how to achieve an export with "everything" included. Sometimes, we faced confusing explanations on checkboxes or other UI elements, e.g., on `salesforce.com` where the checkbox description does not clearly state whether one has to check or uncheck to receive all data. Similarly, we found 8 instances of *Conflicting Information*, such as stating different preferred paths for a DSAR without clarity on the currently intended path.

**Stirring.** Here, we have two types of dark patterns, namely *Emotional Steering* with 22 occurrences, often achieved by intimidating the user by suggesting multiple requests could impose a cost on the user, especially without specifying what number of requests would lead to a charge or in what dimension this charge would lie in, e.g., on `issuu.com` or `tradingview.com`. This is often included in the privacy policy along with information about the user rights. One website, namely `mediafire.com`, also used the privacy policy to assure they are "not in the business of tracking [...]", painting a positive picture about their data processing practices. Some services (e.g., `bloomberg.com`) explicitly state that users exercising their rights might impact the provider's ability to maintain the service. The other 15 cases in the category *Stirring* are of the type *Hidden in Plain Sight*, which is already quite descriptive. Instances of this pattern include `myspace.com` where the request button is greyed out and placed non-prominently or, on `mediafire.com`, where the link to the privacy policy only becomes visible after clicking on a small icon in one of the bottom corners.

**Fickle.** The second least common category *Fickle* consists mostly of the dark pattern *Inconsistent Interface* with 17 occurrences, where user expectations are not met with the usual website structure, e.g., a footer containing a link to the privacy policy only on some of the pages as seen on `salesforce.com`, `linktr.ee`, or `slack.com`. A particular example of *Inconsistent Interface* is deviantart.com, where a GDPR-related menu item is actively hidden based on the URL the user is currently on (Fig. 6c). We found 9 patterns to be a way of *Decontextualizing*, often done by placing data request buttons in unrelated settings, such as in the "Notification Settings" (figma.com). Examples of *Language Discontinuity* (4 occurrences) are showing parts of the privacy policy in a language different from the rest of the privacy policy. In 2 cases, we found interfaces that are *Lacking Hierarchy*, e.g., services that split information about the privacy controls across different products or across different categories, such as unity3d.com.

**Skipping.** The most common dark pattern in *Skipping* is *Look Over There* with 18 cases, where a user is diverted from their primary goal by placing distracting elements concurrently with the desired actions. For example, banners that show up with information about other privacy rights while trying to exercise a data subject right, as encountered on ibm.com. Trying to keep unfavorable options for the user by defaulting to them (*Deceptive Snugness* with 6 examples) was mostly found by services trying to give users incomplete data through defaulting to a short timespan (e.g., facebook.com) for the report of the personal data or pre-selecting only some data categories (e.g., google.com).

We found one particular interesting new dark pattern we placed in the category *Skipping*. Namely, on the website ea.com, one is told to actively check every hour whether the data is now prepared and ready to download, while also giving only 24 h to download instead of sending an e-mail notification. This could lead to users forgetting about it for a sufficient time and then having to re-request the data. According to the description in the taxonomy, such behavior is in the category *Skipping* because the interface seems to be *designed in such a way that users forget* (cf. description in [19]) to actively check sufficiently frequent in the timeframe of 24 h. However, it does not fit into the type of *Deceptive Snugness* as there are no pre-selected default options. Furthermore, the type *Look Over There* is not fitting as no elements compete for the user's attention. One could argue this creates a *Dead End (Obstructing)* for the user since they will not be able to download the data anymore once they forgot to actively check for the readiness of the download (until re-requesting). However, we argue that this constitutes another dark pattern, which only results from the dark pattern we currently have at hand. Namely, to burden the user with the task of actively polling for the download, while the service would have the means to send e-mail notifications. Hence, we name this dark pattern *Active User Polling*. We do not consider this as indicative of a non-fitting taxonomy, as this is quite a specific implementation unlikely to occur frequently.

## 5   Discussion

The systematic examination of 166 popular websites led to 238 instances of identified dark patterns (RQ2) as shown in Fig. 5. This finding would support the claim of service providers strategically introducing barriers on their websites, potentially discouraging users from exercising their right of access. Interestingly, some descriptions found by [59] were observed on a larger scale in our study as well. For instance, *making it impossible to access GDPR requests* [59] often aligns with the dark pattern *Dead End* from the EDPB taxonomy [19], which we encountered multiple times. Within our dataset of popular account-based websites, approximately one-third showed no dark patterns. However, in extreme cases we labeled up to 11 dark patterns on a single website (Fig. 4).

Regarding the types of dark patterns (RQ3), obstructive patterns were the most prevalent, confirming findings similar to those of Kelly and Burkell [34] who categorize *Obstruction* as one of the main categories in their typology. The distribution across the other five categories did not show a significant preference. During the coding phase, placing instances into categories and corresponding types proved straightforward, supporting our assumption that the EDPB taxonomy [19] is appropriate for studies investigating data subject rights. Despite this, we added a new dark pattern, *Active User Polling*, to the existing category *Skipping*, although we suspect this to be a rare implementation.

While the amount of dark patterns could align with Waldman's assumption [71] that the industry is disinterested in meaningful privacy, we also acknowledge that discerning intentional dark patterns from unintentional ones is not straightforward. For example, the hidden menu item in Fig. 6c could be a programming error. Some patterns might even serve a non-malicious purpose. The 60-second timer shown in Fig. 6a could safeguard against denial-of-service attack. Note that this uncertainty is inherent in the nature of dark patterns. Consequently, we provide the codebook including notes and screencasts to make each case transparent. Nevertheless, the observed increase in time with a higher number of dark patterns (Fig. 4) suggests hindrances in posing DSARs on such websites, potentially infringing on *Art. 12 para. 2* of the GDPR. This aligns with previous findings [10,70] which show at least questionable usability of DSAR submission processes. Such shortcomings can induce a negative view of data protection policies for end users [5] and distrust in providers [10]. This may even lead users to abstain from exercising their rights [62,67].

Similar to prior studies [3,68], we found that various websites offer multiple DSAR submission mechanisms (RQ2), with e-mail addresses and website forms being the most common (see Fig. 3). Direct data download buttons are scarce, suggesting manual DSAR responses over automated ones. Furthermore, we see there are four websites where all three researchers struggled to find adequate information, suggesting a lack of information or even intentional concealment. While the number of such cases seems low compared to [3], it is noteworthy that each of the three researchers searched for a mechanism and only in four cases, none of them could find enough information. Furthermore, our corpus consists of highly popular websites that presumably try to adhere to different regional data policies nowadays and thus often provide at least an e-mail address for DSARs.

## 5.1  Ethical Considerations

In this study, we did not deal with the personal data of other users. Rather, we used freshly generated anonymous e-mail accounts and fictitious data. Nonetheless, online services might have stored secondary data, e.g., IP addresses. To authenticate to such data, we relied on account credentials, assuming any data potentially returned is associated with our test accounts. Additionally, we limited the dataset to websites that we could investigate without requiring privacy-sensitive information. We introduce as little manual work and expenses for data protection officers of companies as possible by restricting our actions to the website interfaces. Typically, this meant refraining from following up on automated e-mails requesting DSAR submission confirmation. In a few other cases, the response was anyway automated and gave a direct download link and hence did not generate additional manual work.

Note that we decided against notifying the 113 website operators about the dark patterns detailed in our study. Responsible disclosure, common in vulnerability reporting, involves notifying responsible actors to fix issues before public disclosure to prevent malicious exploitation by third parties. However, none of the identified dark patterns allowed for third-party exploitation. The potential impact on company reputation was also deemed low, given the extensive documentation of dark pattern use across various domains such as shopping [50], consent notifications [35, 38, 55], legitimate interest [40] and even in the case of the right of access [43, 59], with some already explicitly naming companies. Hence, we think the reputational impact is marginal at most. Nevertheless, we would welcome an effect on stricter enforcement of regional data regulations such as the GDPR or even regular audits for dark patterns as proposed by [44]. Due to the inherent plausible deniability of dark patterns, we also refrain from seeking statements from website operators, since we have no reliable way to verify the veracity of such statements. Reporting only trends or aggregated numbers would compromise the transparency and reproducibility of our work by keeping codebooks and screencasts unpublished. We think transparent publication still allows willing companies to address unintentional dark patterns without significant reputational risk.

## 5.2  Limitations

Our study was conducted through manual analysis, thus it is possible that we did not encounter every dark pattern implemented on a website. The search protocol used might have excluded request paths, leading to reported dark pattern numbers possibly representing the lower boundary of actual occurrences. As already discussed, we can never be sure that a dark pattern was intentional. We tried to counter the ambiguity induced by incorrectly labeled patterns by having three researchers with different backgrounds (security and privacy, human-computer interaction, and technical communication) conduct the analysis and discuss all the results until consensus. Note that, we cannot provide inter-rater agreement scores as we started with an independent open coding phase without deciding on dark pattern types from the taxonomy.

Another point to consider is that we might have gotten familiar with common processes or frequently employed dark patterns. This could lead to distortions in the timing measurements and to overlooking of instances of dark patterns. While the three researchers who conducted the analysis have different backgrounds, all three bring expertise in their respective field, possibly influencing the results.

We focused solely on the websites of online services, while there is no specific manner prescribed on how data subjects have to pose a DSAR. Additionally, we concentrated on the DSAR submission process, omitting the recording or examination of responses. Existing work, such as [10,39,59,68,70], already show a trend of low usability for end users in data controller responses. We intentionally made no statements about the conformity of singular webpages, as this needs to be decided on a case-by-case basis. We can merely supply evidence for or against the conformity. Additionally, there are other data subject rights which may face similar hindrances from deceptive design choices. Investigating different data subject rights helps to enhance the understanding of the practical implementations of data protection regulations. While our work shows a trend in dark pattern usage among popular websites, further research with larger datasets is essential to strengthen such findings. Finally, our assessments were conducted exclusively on desktop versions of the websites. The dark pattern landscape may differ on mobile versions or on apps. Kröger et al. [39] have highlighted numerous issues with the right of access on apps. Hence, it is not too far off to suspect dark patterns on mobile platforms.

## 6   Conclusion

We analyzed 166 of the 500 most popular websites for their use of dark patterns inhibiting DSARs and found that on 113 of 166, i.e., 68%, websites contain at least one dark pattern, and often more than one, totaling 238 instances of dark patterns. We used the taxonomy given by the EDPB [19] to categorize our findings, linking related GDPR articles. We found that the most common patterns are of obstructing nature and showed that higher numbers of dark patterns correlate with longer times to submit a DSAR. Our work adds to documenting dark patterns in various areas, highlighting their presence also in the realm of data subject rights. Our work shows that despite the ongoing efforts of policymakers to grant users more digital rights, especially concerning the processing of personal data, there are still impediments to overcome before users are able to achieve easy-to-use data transparency. While it remains unclear whether the hindrances are always intentionally, users refrain in large parts from exercising their rights [62] and we suspect deceptive design choices to contribute to this phenomen. Another point to raise is that while policymakers actively work on legislation covering dark patterns (such as the *Digital Services Act* [22]), regulations should be enforced, e.g., through regular audits. For developers, possibly even unaware of the deceptive nature of implemented designs, another interesting direction is the growing research trend of "fair patterns" [60]. Concluding, there is still a lot of potential to improve on to enable end users to make use of their digital rights.

## 6.1 Future Work

Future steps should expand measurements to validate findings on a larger scale. To enable these, it is necessary to investigate the feasibility of automating such dark pattern checks, as manual inspection becomes impractical with a more extensive data set. This could even facilitate periodic measurements to evaluate the impact of new regulations. Assessments of dark patterns hindering data subject rights could also extend to platforms other than desktop websites, such as mobile browsers or apps. As there are more data subject rights than the right of access, it can also be suitable to explore dark patterns in relation to these rights. Researchers should examine different countermeasures and regulations against dark patterns, and keep investigating the effects of new regulations.

**Acknowledgments.** A. Löbel was supported by the research project "North-Rhine Westphalian Experts in Research on Digitalization (NERD II)", sponsored by the state of North Rhine-Westphalia – NERD II 005-2201-0014. The work of R. Schäfer was funded in part by the German B-IT Foundation. The work of H. Püschel was supported by the PhD School "SecHuman - Security for Humans in Cyberspace" by the federal state of NRW, Germany.

# References

1. Access Now, Simply Secure, World Wide Web Foundation: Comments to the EDPB consultation on Guidelines 3/2022 on dark patterns in social media platform interfaces (2023)
2. Alizadeh, F., Jakobi, T., Boldt, J., Stevens, G.: GDPR-reality check on the right to access data: claiming and investigating personally identifiable data from companies. Mensch Und Computer (2019). https://doi.org/10.1145/3340764.3344913
3. Ausloos, J., Dewitte, P.: Shattering one-way mirrors. Data subject access rights in pactice. Int. Data Privacy Law **8**(1), 4–28 (2018). https://doi.org/10.1093/idpl/ipy001
4. Bollinger, D., Kubícek, K., Jiménez, C.C., Basin, D.A.: Automating cookie consent and GDPR violation detection. In: USENIX Security Symposium (2022). https://www.usenix.org/conference/usenixsecurity22/presentation/bollinger
5. Bongard-Blanchy, K., Rossi, A., Rivas, S., Doublet, S., Koenig, V., Lenzini, G.: I am definitely manipulated, even when I am aware of it. It's ridiculous!" - Dark patterns from the end-user perspective. In: Designing Interactive Systems Conference (2021). https://doi.org/10.1145/3461778.3462086
6. Boniface, C., Fouad, I., Bielova, N., Lauradoux, C., Santos, C.: Security analysis of subject access request procedures. In: Naldi, M., Italiano, G.F., Rannenberg, K., Medina, M., Bourka, A. (eds.) APF 2019. LNCS, vol. 11498, pp. 182–209. Springer, Cham (2019). https://doi.org/10.1007/978-3-030-21752-5_12
7. Borberg, I., Hougaard, R., Rafnsson, W., Kulyk, O.: "So I Sold My Soul": effects of dark patterns in cookie notices on end-user behavior and perceptions. In: Symposium on Usable Security (2022). https://doi.org/10.14722/usec.2022.23026

8. Bösch, C., Erb, B., Kargl, F., Kopp, H., Pfattheicher, S.: Tales from the dark side: privacy dark strategies and privacy dark patterns. Privacy Enhanc. Technol. (2016). https://doi.org/10.1515/popets-2016-0038

9. Botes, W.M., Carli, R., Rossi, A., Sanchez Chamorro, L., Santos, C., Sergeeva, A.: Feedback to the Guidelines 3/2022 on "Dark patterns in social media platform interfaces: How to recognise and avoid them" (2022)

10. Bowyer, A., Holt, J., Go Jefferies, J., Wilson, R., Kirk, D., David Smeddinck, J.: Human-GDPR interaction: practical experiences of accessing personal data. In: Conference on Human Factors in Computing Systems (2022). https://doi.org/10.1145/3491102.3501947

11. Bufalieri, L., Morgia, M.L., Mei, A., Stefa, J.: GDPR: when the right to access personal data becomes a threat. In: IEEE International Conference on Web Services (2020). https://doi.org/10.1109/ICWS49710.2020.00017

12. Bygrave, L.A.: Data Privacy Law: An International Perspective (2014). https://doi.org/10.1093/acprof:oso/9780199675555.001.0001

13. Cagnazzo, M., Holz, T., Pohlmann, N.: GDPiRated – stealing personal information on- and offline. In: Sako, K., Schneider, S., Ryan, P.Y.A. (eds.) ESORICS 2019. LNCS, vol. 11736, pp. 367–386. Springer, Cham (2019). https://doi.org/10.1007/978-3-030-29962-0_18

14. California Consumer Privacy Act of 2018 (2018)

15. Decision of the EEA Joint Committee No. 154/2018 of July 6, 2018 (2018)

16. Di Geronimo, L., Braz, L., Fregnan, E., Palomba, F., Bacchelli, A.: UI dark patterns and where to find them: a study on mobile applications and user perception. In: Conference on Human Factors in Computing Systems (2020). https://doi.org/10.1145/3313831.3376600

17. Di Martino, M., Meers, I., Quax, P., Andries, K., Lamotte, W.: Revisiting identification issues in GDPR 'Right Of Access' Policies: a technical and longitudinal analysis. Privacy Enhanc. Technol. **2022**(2), 95–113 (2022). https://doi.org/10.2478/popets-2022-0037

18. Lupiáñez-Villanueva, F., Boluda, A., Bogliacino, F., Liva, G., Lechardoy, L., Rodríguez de las Heras Ballell, T.: Behavioural study on unfair commercial practices in the digital environment - Dark patterns and manipulative personalisation - Final Report. In: European Commission, Directorate-General for Justice and Consumers (2022). https://doi.org/10.2838/859030

19. European Data Protection Board. Guidelines 3/2022 on Dark patterns in social media platform interfaces: how to recognise and avoid them (2022)

20. European Parliament, Council of the European Union: Directive 2009/136/EC of the European Parliament and of the Council

21. European Parliament, Council of the European Union. Regulation (EU) 2016/679 of the European Parliament and of the Council (2016)

22. European Parliament, Council of the European Union. Regulation (EU) 2022/2065 of the European Parliament and of the Council (2022)

23. Flick, U.: An Introduction to Qualitative Research (2022)

24. Gray, C.M., Santos, C., Bielova, N.: Towards a preliminary ontology of dark patterns knowledge. In: Extended Abstracts of the Conference on Human Factors in Computing Systems (2023). https://doi.org/10.1145/3544549.3585676

25. Gray, C.M., Santos, C.T., Bielova, N., Mildner, T.: An ontology of dark patterns knowledge: foundations, definitions, and a pathway for shared knowledge-building. In: CHI Conference on Human Factors in Computing Systems (2024). https://doi.org/10.1145/3613904.3642436

26. Greenleaf, G.: Global tables of data privacy laws and bills. Privacy Laws Bus. Int. Rep. (2021). https://doi.org/10.2139/ssrn.3836261

27. Gunawan, J., Pradeep, A., Choffnes, D., Hartzog, W., Wilson, C.: A comparative study of dark patterns across web and mobile modalities. ACM Hum. Comput. Interact. 5(CSCW2), 1–29 (2021). https://doi.org/10.1145/3479521

28. Gundelach, R., Herrmann, D.: Cookiescanner: an automated tool for detecting and evaluating GDPR consent notices on websites. In: International Conference on Availability, Reliability and Security (2023). https://doi.org/10.1145/3600160.3605000

29. Habib, H., Li, M., Young, E., Cranor, L.: "Okay, whatever": an evaluation of cookie consent interfaces. In: Conference on Human Factors in Computing Systems (2022). https://doi.org/10.1145/3491102.3501985

30. Habib, H., et al.: "It's a Scavenger Hunt": usability of websites' opt-out and data deletion choices. In: Conference on Human Factors in Computing Systems (2020). https://doi.org/10.1145/3313831.3376511

31. Hennemann, M., Lienemann, G., Sprikl, C.: Mapping Global Data Law. University of Passau Institute for Law of the Digital Society Research Paper (2022)

32. Hidaka, S., Kobuki, S., Watanabe, M., Seaborn, K.: Linguistic dead-ends and alphabet soup: finding dark patterns in Japanese apps. In: Conference on Human Factors in Computing Systems (2023). https://doi.org/10.1145/3544548.3580942

33. Jarovsky, L.: Dark patterns in personal data collection: definition. Taxonomy Lawfulness (2022). https://doi.org/10.2139/ssrn.4048582

34. Kelly, D., Burkell, J.: Documenting Privacy Dark Patterns: How Social Networking Sites Influence Users' Privacy Choices, vol. 376. FIMS Publications (2023)

35. Kirkman, D., Vaniea, K., Woods, D.W.: DarkDialogs: automated detection of 10 dark patterns on cookie dialogs. In: IEEE European Symposium on Security and Privacy (2023). https://doi.org/10.1109/EuroSP57164.2023.00055

36. Kowalczyk, M., Gunawan, J.T., Choffnes, D., Dubois, D.J., Hartzog, W., Wilson, C.: Understanding dark patterns in home IoT devices. In: Conference on Human Factors in Computing Systems (2023). https://doi.org/10.1145/3544548.3581432

37. Kretschmer, M., Pennekamp, J., Wehrle, K.: Cookie banners and privacy policies: measuring the impact of the GDPR on the web. ACM Trans. Web 15(4), 1–42 (2021). https://doi.org/10.1145/3466722

38. Krisam, C., Dietmann, H., Volkamer, M., Kulyk, O.: Dark patterns in the wild: review of cookie disclaimer designs on top 500 German websites. In: European Symposium on Usable Security (2021). https://doi.org/10.1145/3481357.3481516

39. Kröger, J.L., Lindemann, J., Herrmann, D.: How do app vendors respond to subject access requests? A longitudinal privacy study on iOS and Android Apps. In: International Conference on Availability, Reliability and Security (2020). https://doi.org/10.1145/3407023.3407057

40. Kyi, L., Ammanaghatta Shivakumar, S., Santos, C.T., Roesner, F., Zufall, F., Biega, A.J.: Investigating deceptive design in GDPR's legitimate interest. In: Conference on Human Factors in Computing Systems (2023). https://doi.org/10.1145/3544548.3580637

41. Lauradoux, C.: Can authoritative governments abuse the right to access? In: Gryszczyńska, A., Polański, P., Gruschka, N., Rannenberg, K., Adamczyk, M. (eds.) APF 2022. LNCS, pp. 23–33. Springer, Cham (2022). https://doi.org/10.1007/978-3-031-07315-1_2

42. Le Pochat, V., Van Goethem, T., Tajalizadehkhoob, S., Korczyński, M., Joosen, W.: Tranco: a research-oriented top sites ranking hardened against manipulation.

In: Annual Network and Distributed System Security Symposium (2019). https://doi.org/10.14722/ndss.2019.23386

43. Li, W., Li, Z., Li, W., Zhang, Y., Li, A.: Mapping the empirical evidence of the GDPR (In-)Effectiveness: a systematic review (2023)

44. Luguri, J., Strahilevitz, L.J.: Shining a light on dark patterns. J. Legal Anal. **13**(1), 43–109 (2021). https://doi.org/10.1093/jla/laaa006

45. Machuletz, D., Böhme, R.: Multiple purposes, multiple problems: a user study of consent dialogs after GDPR. Privacy Enhanc. Technol. **2019**(2), 481–498 (2019). https://doi.org/10.2478/popets-2020-0037

46. Mahieu, R., Asghari, H., van Eeten, M.: Collectively exercising the right of access: individual effort, societal effect. Internet Policy Rev. **7**(3) (2018). https://doi.org/10.14763/2018.3.927

47. Mahieu, R., Asghari, H., Parsons, C., van Hoboken, J., Crete-Nishihata, M., Hilts, A., Anstis, S.: Measuring the brussels effect through access requests: has the European general data protection regulation influenced the data protection rights of Canadian citizens? J. Inf. Policy **11**, 301–349 (2021). https://doi.org/10.5325/jinfopoli.11.2021.0301

48. Mahieu, R.: The right of access to personal data: a genealogy. Technol. Regulat. **2021** (2021). https://doi.org/10.26116/techreg.2021.005

49. Martino, M.D., Robyns, P., Weyts, W., Quax, P., Lamotte, W., Andries, K.: Personal information leakage by abusing the GDPR 'Right of Access'. In: Symposium on Usable Privacy and Security (2019). https://www.usenix.org/conference/soups2019/presentation/dimartino

50. Mathur, A., et al.: Dark patterns at scale: findings from a crawl of 11K shopping websites. ACM Hum. Comput. Interact. **3**(CSCW) (2019). https://doi.org/10.1145/3359183

51. Mathur, A., Kshirsagar, M., Mayer, J.: What makes a dark pattern... dark? Design attributes, normative considerations, and measurement methods. In: Conference on Human Factors in Computing Systems (2021). https://doi.org/10.1145/3411764.3445610

52. Mildner, T., Savino, G.L., Doyle, P.R., Cowan, B.R., Malaka, R.: About engaging and governing strategies: a thematic analysis of dark patterns in social networking services. In: Conference on Human Factors in Computing Systems (2023). https://doi.org/10.1145/3544548.3580695

53. Monge Roffarello, A., Lukoff, K., De Russis, L.: Defining and identifying attention capture deceptive designs in digital interfaces. In: Conference on Human Factors in Computing Systems (2023). https://doi.org/10.1145/3544548.3580729

54. Norris, C., De Hert, P., L'hoiry, X., Galetta, A.: The unaccountable state of surveillance. Exercising Access Rights in Europe. https://doi.org/10.1007/978-3-319-47573-8

55. Nouwens, M., Liccardi, I., Veale, M., Karger, D., Kagal, L.: Dark patterns after the GDPR: scraping consent pop-ups and demonstrating their influence. In: Conference on Human Factors in Computing Systems (2020). https://doi.org/10.1145/3313831.3376321

56. Pavur, J., Knerr, C.: GDPArrrrr: Using Privacy Laws to Steal Identities (2019)

57. Petelka, J., Oreglia, E., Finn, M., Srinivasan, J.: Generating practices: investigations into the double embedding of GDPR and data access policies. ACM Hum. Comput. Interact. **6**(CSCW2) (2022). https://doi.org/10.1145/3555631

58. Pins, D., Jakobi, T., Stevens, G., Alizadeh, F., Krüger, J.: Finding, getting and understanding: the user journey for the GDPR's right to access. Behav. Inf. Technol. **41**(10) (2022). https://doi.org/10.1080/0144929X.2022.2074894

59. Pöhn, D., Mörsdorf, N., Hommel, W.: Needle in the haystack: analyzing the right of access according to GDPR article 15 five years after the implementation. In: International Conference on Availability, Reliability and Security (2023). https://doi.org/10.1145/3600160.3605064

60. Potel-Saville, M., Da Rocha, M.: From dark patterns to fair patterns? Usable taxonomy to contribute solving the issue with countermeasures. In: Rannenberg, K., Drogkaris, P., Lauradoux, C. (eds.) Privacy Technologies and Policy. Springer, Cham (2024). https://doi.org/10.1007/978-3-031-61089-9_7

61. Raento, M.: The data subject's right of access and to be informed in Finland: an experimental study. Int. J. Law Inf. Technol. **14**(3) (2006). https://doi.org/10.1093/ijlit/eal008

62. Rughiniş, R., Rughiniş, C., Vulpe, S.N., Rosner, D.: From social netizens to data citizens: variations of GDPR awareness in 28 European countries. Comput. Law Secur. Rev. **42**, 10558 (2021). https://doi.org/10.1016/j.clsr.2021.105585

63. Schade, F.: Dark sides of data transparency: organized immaturity after GDPR? Bus. Ethics Quart. **33**(3) (2023). https://doi.org/10.1017/beq.2022.30

64. Schäfer, R., Preuschoff, P.M., Röpke, R., Sahabi, S., Borchers, J.: Fighting malicious designs: towards visual countermeasures against dark patterns. In: Conference on Human Factors in Computing Systems (2024). https://doi.org/10.1145/3613904.3642661

65. Singh, J., Cobbe, J.: The security implications of data subject rights. IEEE Secur. Privacy **17**(6) (2019). https://doi.org/10.1109/MSEC.2019.2914614

66. Soe, T.H., Nordberg, O.E., Guribye, F., Slavkovik, M.: Circumvention by design - dark patterns in cookie consent for online news outlets. In: Nordic Conference on Human-Computer Interaction: Shaping Experiences, Shaping Society (2020). https://doi.org/10.1145/3419249.3420132

67. Sørum, H., Presthus, W.: Dude, where's my data? The GDPR in practice, from a consumer's point of view. Inf. Technol. People **34**(3) (2021). https://doi.org/10.1108/ITP-08-2019-0433

68. Urban, T., Tatang, D., Degeling, M., Holz, T., Pohlmann, N.: A study on subject data access in online advertising after the GDPR. In: Pérez-Solà, C., Navarro-Arribas, G., Biryukov, A., Garcia-Alfaro, J. (eds.) DPM/CBT -2019. LNCS, vol. 11737, pp. 61–79. Springer, Cham (2019). https://doi.org/10.1007/978-3-030-31500-9_5

69. Utz, C., Degeling, M., Fahl, S., Schaub, F., Holz, T.: (Un)Informed consent: studying GDPR consent notices in the field. In: ACM SIGSAC Conference on Computer and Communications Security (2019). https://doi.org/10.1145/3319535.3354212

70. Veys, S., et al.: Pursuing usable and useful data downloads under GDPR/CCPA access rights via co-design. In: Symposium on Usable Privacy and Security (2021). https://www.usenix.org/conference/soups2021/presentation/veys

71. Waldman, A.E.: Industry Unbound: The Inside Story of Privacy, Data, and Corporate Power (2021). https://doi.org/10.1017/9781108591386

72. Younas, A., Ogli Mirzaraimov, B.T.: To what extent are consumers harmed in the digital market from the perspective of the GDPR? Int. J. Multidiscip. Res. Anal. **4**(8) (2021). https://doi.org/10.47191/ijmra/v4-i8-17

73. Zagal, J.P., Björk, S., Lewis, C.: Dark patterns in the design of games. In: Foundations of Digital Games (2013)

# AI Cards: Towards an Applied Framework for Machine-Readable AI and Risk Documentation Inspired by the EU AI Act

Delaram Golpayegani[1]([envelope]) [iD], Isabelle Hupont[2][iD], Cecilia Panigutti[2][iD], Harshvardhan J. Pandit[3][iD], Sven Schade[2][iD], Declan O'Sullivan[1][iD], and Dave Lewis[1][iD]

[1] ADAPT Centre, Trinity College Dublin, Dublin, Ireland
{sgolpays,declan.osullivan,delewis}@tcd.ie
[2] European Commission, Joint Research Centre (JRC), Ispra, Italy
[3] ADAPT Centre, Dublin City University, Dublin, Ireland
me@harshp.com

**Abstract.** With the upcoming enforcement of the EU AI Act, documentation of high-risk AI systems and their risk management information will become a legal requirement playing a pivotal role in demonstration of compliance. Despite its importance, there is a lack of standards and guidelines to assist with drawing up AI and risk documentation aligned with the AI Act. This paper aims to address this gap by providing an in-depth analysis of the AI Act's provisions regarding technical documentation, wherein we particularly focus on AI risk management. On the basis of this analysis, we propose **AI Cards** as a novel holistic framework for representing a given intended use of an AI system by encompassing information regarding technical specifications, context of use, and risk management, both in human- and machine-readable formats. While the human-readable representation of AI Cards provides AI stakeholders with a transparent and comprehensible overview of the AI use case, its machine-readable specification leverages on state of the art Semantic Web technologies to embody the interoperability needed for exchanging documentation within the AI value chain. This brings the flexibility required for reflecting changes applied to the AI system and its context, provides the scalability needed to accommodate potential amendments to legal requirements, and enables development of automated tools to assist with legal compliance and conformity assessment tasks. To solidify the benefits, we provide an exemplar AI Card for an AI-based student proctoring system and further discuss its potential applications within and beyond the context of the AI Act.

**Keywords:** AI documentation · Risk management · EU AI Act · Semantic Web

M. Jensen et al. (Eds.): APF 2024, LNCS 14831, pp. 48–72, 2024.
https://doi.org/10.1007/978-3-031-68024-3_3

# 1    Introduction

Artificial Intelligence (AI) has become a centrepiece of reshaping many aspects of individual life, society, and public and private sectors for the better [10,19,28]. Yet, concerns have been raised regarding the ethical implications and potential risks associated with the use of AI, such as biases in decision-making algorithms [8,31] and privacy issues [25,47]. These concerns have prompted a global wave of trustworthy AI guidelines [14,35] and some preliminary regulatory efforts aimed at mitigating the potential harms of AI, ensuring its development and use are aligned with safety standards and respect human rights and freedoms. The European Union's (EU) AI Act [2][1] is the first comprehensive legal framework on AI, through which AI systems are subjected to a set of regulatory obligations according to their risk level-that can be, from highest to lowest, *unacceptable, high-risk, limited-risk,* or *minimal-risk.*

With the risk level being the yardstick for determining the regulatory requirements that AI systems need to satisfy, the AI Act is intent on promoting secure, trustworthy, and ethical use of AI, with a particular emphasis on management and mitigation of potential harms that high-risk AI systems may cause. The risk management system provisions, set in Art. 9, thereby play a pivotal role in the implementation of the Act. Effective AI risk management requires information regarding the AI system, its incorporating components, its context of use, and its potential risks to be maintained and communicated in the form of *documented information.* While maintaining and sharing information regarding AI systems and their risks promote transparency and in turn trustworthiness [24], it is additionally a legal obligation for providers of high-risk AI systems to draw up technical documentation to demonstrate compliance with the Act's requirements (Art. 11). The elements of technical documentation, which includes risk management system, are described in Annex IV at a high-level. However, to serve its purpose technical documentation needs to be extensive and detailed. Being limited to defining essential requirements, the AI Act relies on European *harmonised standards* for technical specifications to help with implementation and enforcement of its legal requirements, including technical documentation [29,45]. Notably, the European Commission's draft standardisation request [12] refrains from calling for European standards in relation to data, model, or AI system documentation [18]. Consequently, the European Standardisation Organisations' work plan does not include standards specifically addressing these aspects [44]. In parallel, the existing body of work on AI, model, and data documentation approaches pays little attention to documentation for legal compliance, fails to provide a strong connection between the documented information and responsible AI implications, and does not take into consideration aspects related to risk management [21,24].

---

[1] For this work, we examined multiple AI Act mandates published since April 2021, in particular the agreed provisional text. However, the references to the AI Act within this paper shall be interpreted as references to the European Commission's proposal; as at the time of writing, the final content of the AI Act is not published in the official journal of the European Union.

Currently, there is a critical need for technical documentation to support and be in sync with the AI system development and usage practices. Moreover, with the involvement of several entities across the AI value chain, the documentation must be generated and maintained in a manner that ensures consistency and interoperability. This is also crucial for investigation of technical documentation by AI auditors and conformity assessment bodies for compliance checking and certification. The motivation for this work is therefore to address this current need for consistent, uniform, and interoperable specifications to support effective implementation of the AI Act. This paper aims to address the current lack of unified technical documentation practices aligned with the AI Act, with a threefold contribution:

1. we provide an in-depth analysis of the provisions of the EU AI Act in regard to documentation with a focus on *technical documentation* and *risk management system documentation* (Sect. 3);
2. we propose *AI Cards*-a novel framework providing a human-readable overview of a given use of an AI system and its risks (Sect. 4); and
3. we present a machine-readable specification for AI Cards that enables generating, maintaining, and updating documentation in sync with AI development. It further supports querying and sharing information needed for tasks such as compliance checking, comparing multiple AI specifications for purposes such as AI procurement, and exchanging information across the value chain (Sect. 4.3).

We demonstrate a practical application of AI Cards using an illustrative example of an AI-based proctoring system (Sect. 5), validate its usefulness through a survey (Sect. 6), and discuss its benefits (Sect. 7).

## 2   Related Work

### 2.1   Previous Studies on the AI Act's Documentation and Risk Management Requirements

Since the release of the AI Act's proposal by the European Commission in April 2021, the research community, international organisations, and industrial actors have been exploring the new avenues it opened by analysing the AI Act's contents for suggesting clarifications, identifying potential gaps, or giving critique. In this section, we refer to some existing studies on the provisions of the AI Act in regard to documentation and risk management.

Panigutti et al. explore the role of the AI Act's transparency and documentation requirements in addressing the opacity of high-risk AI systems [39]. Gyevnara et al. discuss compliance-oriented transparency required to satisfy the AI Act's requirements in regard to risk and quality management systems [20]. Schuett presents a comprehensive analysis of the AI Act's risk management provisions, without delving into the details of risk management system documentation [43]. Soler et al.'s assessment of international standards in regard to addressing AI Act's requirements shows that these standards are insufficient to meet the provisions of the AI Act related to risk management [44].

In regard to technical documentation (Art. 11 and Annex IV), the most comprehensive studies, to the best of our knowledge, are the work of Hupont et al. [24], which identifies 20 information elements needed in documentation of AI systems and their constituting datasets, and the analysis of Annex IV provided by Golpayegani et al. [17], which identifies 50 information elements. Nevertheless, both studies remain short in being fully comprehensive in terms of covering both technical and risk management system documentation requirements laid down in Art. 11, Annex IV, and Art. 9. Building on our previous work, presented in [17] and [24], this paper provides an in-depth analysis of technical documentation with a focus on risk management system documentation.

## 2.2 Alignment of Existing Documentation Practices with the AI Act

Compliance with the AI Act's documentation requirements requires guidelines, standards, tools, and formats to assist with generation, communication, and auditing of technical documentation. With transparency being widely recognised as a key factor in implementing trustworthy AI [14], several documentation frameworks proposed by the AI community, such as Datasheets for Datasets [16] and Model Cards [32], have become *de-facto* practices. As these documentation approaches are widely-adopted, a key question is the extent to which they could be leveraged for regulatory compliance tasks. This question is investigated by the studies mentioned in the following. Pistilli et al. discuss the potential of Model Cards as a compliance tool and anticipate its adoption-among other existing documentation practices originated from the AI community-for compliance with the AI Act's documentation obligations [40]. The work by Hupont et al. [24], which investigates the 6 most widely-used AI and data documentation approaches for their alignment with the implementation of documentation provisions, concludes that AI Factsheets [9] offers a higher overall degree of information coverage, followed by Model Cards and the AI Classification Framework proposed by the Organization for Economic Cooperation and Development (OECD) [35]. The research also demonstrates that while data-related information elements are well-covered by most documentation approaches, particularly Datasheets for Datasets [16], the Dataset Nutrition Label [22], and the Accountability for Machine Learning framework [26], they still do not cover technical information needs related to AI systems. This finding is further strengthened in a follow-up comparative analysis of 36 AI system, model, and/or dataset documentation practices, which unravels the overall alignment of documentation practices with transparency requirements of the EU AI Act and other recent EU data and AI initiatives, and spots a gap in representing the information related to AI systems in its entirety and its context of use [30]. Interestingly, to date, the only documentation methodology conceived from design for the AI Act is *Use Case Cards* [23], which primarily focuses on documenting the intended use of AI systems to assess their risk level as per the AI Act, without mentioning technical information and risk management details.

### 2.3   Machine-Readable Data, Model, and AI System Documentation

While documentation approaches are acknowledged as instruments for improving transparency, and in turn enhancing trustworthiness, there has been little attention to the barriers in generation, maintenance, assessment, and exchange of conventional text-based documentation. Providing machine-readable specifications is an idea taken up by some recent work to support adaptable and interoperable documentation. In the area of data documentation, Open DataSheets[2], proposed by Roman et al., provides a metadata framework for documenting open datasets in a machine-readable manner [41]. For machine-understandable documentation of machine learning models, Model Card Report Ontology (MCRO)[3], developed by Amith et al., offers the metadata for the content of Model Card reports [7]. Linked Model and Data Cards (LMDC) present a schema for integration of Model and Data Cards in a data space to provide a holistic view of a model or an AI service [13]. The Realising Accountable Intelligent Systems (RAInS) ontology[4], created by Naja et al., models the information relevant to AI accountability traces [33].

In the context of documentation for regulatory compliance, the W3C Data Privacy Vocabulary (DPV) [37][5] is developed based on the requirements of the EU GDPR (General Data Protection Regulation) and has been applied for representing Data Protection Impact Assessment (DPIA) information [36], documenting data breach reports [38], and representing Register of Processing Activities (ROPA) [42]. In our previous work, we proposed the AI Risk Ontology (AIRO)[6] for describing AI systems and their risks based on the AI Act and ISO 31000 family of standards [17]. In this work, AIRO is employed to provide the basis for the machine-readable representation of the AI Cards. However, it is important to note that ISO 31000 family of standards is not aligned with the AI Act due to some fundamental differences [44]. AIRO is used while awaiting future standards to be developed in response to the European Commission's standardisation request. Therefore, it needs to be updated with future publication of harmonised standards related to AI risk management.

## 3   Provisions of the EU AI Act Regarding Documentation of AI Systems

Documentation of AI systems is highly tied to the principles of transparency and accountability [23,33], providing an appropriate degree of information to

---

[2] https://github.com/microsoft/opendatasheets-framework.
[3] https://github.com/UTHealth-Ontology/MCRO.
[4] https://w3id.org/rains.
[5] https://w3id.org/dpv/.
[6] https://w3id.org/airo.

different stakeholders to enable several types of assessments. In a similar vein, documentation is essential for assessing legal compliance. Figure 1 shows the AI Act's documentation requirements for high-risk AI systems and illustrates how they are related to each other. Within the AI Act, *technical documentation* and *quality management system documentation* are key documents in the conformity assessment procedure, containing detailed information required for self- and third-party assessments. Given the central role of technical documentation and presence of related guidance on its content (Annex IV), this work focuses on identifying its concrete information elements. However, among its incorporating documents, our analysis expands on *risk management system documentation* considering that the risk-centred nature of the AI Act makes the risk management system requirement (Art. 9) taking the lead in ensuring that the potential risks of high-risk AI systems are reduced to an acceptable level. In the following, we focus on identifying the information elements required to be featured in technical (Sect. 3.1) and risk management (Sect. 3.2) documentation.

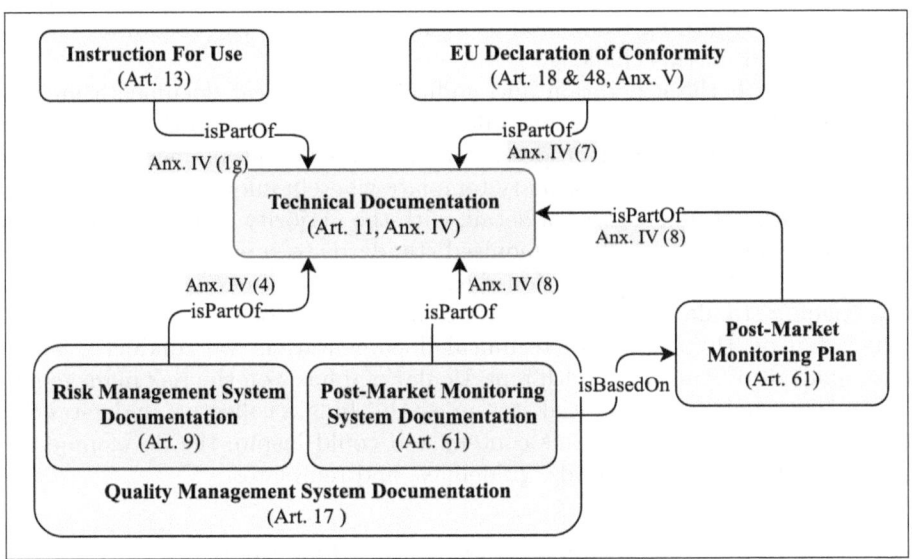

**Fig. 1.** High-risk AI systems' documentation requirements as per the AI Act.

### 3.1   Article 11 and Annex IV - Technical Documentation

As discussed earlier, technical documentation is a key artefact for demonstrating compliance with the AI Act. According to Art. 11, it is the *high-risk AI provider* who should draw up the technical documentation *ex-ante* and keep it *updated*. From this obligation, three critical challenges arise: (i) when development of AI systems includes integrating third-party components into the system, e.g. training data and pre-trained models, high-risk AI providers need to dispense details regarding these components in technical documentation, e.g. information about how data is labelled (Annex IV(2)(d)). Thus, generating technical documentation to fulfill the requirements of the AI Act can become a collective activity, which requires communication and exchange of often sensitive and confidential information. This opens up a myriad of questions concerning trust, accountability, and liability, that lie beyond the scope of this discussion. (ii) Real-world testing environments and sandboxes assist in determining what might be difficult to ascertain ex-ante. However, specificities of the *context of use*, which significantly influence AI risks and impacts, might not be revealed until the system's deployment. This suggests a need for ex-ante involvement of potential users and ex-post feedback loops. (iii) Finally, keeping technical documentation up-to-date and re-assessing legal compliance are hard-to-manage tasks. A key question here is "what changes do trigger the update process?". A conspicuous, albeit not comprehensive, answer is *substantial modifications*-changes that affect the system's compliance with the Act or its intended purpose (Art. 3(23)). However, lack of clarity regarding what modifications are deemed as substantial adds to the complexities of this challenge.

To help with the generation and auditing of technical documentation, the minimum set of its incorporating information elements is outlined in Annex IV, which is subject to the European Commission's potential amendments (Art. 11(3)). Annex IV serves as a primary template wherein information elements are described with varying degrees of detail, with the majority articulated at a high-level. This implies a need for harmonised standards to support the implementation of Art. 11. However, as mentioned earlier, to the best of our knowledge there is no ongoing standardisation activities to address this need. Aiming to provide clarification on the content of technical documentation, we conducted an in-depth analysis of Annex IV, with a particular emphasis on the risk management system. The analysis is an initial step in establishing a collective understanding of the technical documentation's content and could inspire the development of prospective European standards, guidelines, and templates.

### 3.2    Article 9 - Risk Management System Documentation

The AI Act will rely on harmonised standards to operationalise legal requirements at the technical level, and risk management system will be no exception. At the time of writing, however, European Standardisation Organisations are still working on deliverables planned in response to the European Commission's standardisation request. Furthermore, existing ISO/IEC standards on AI risk management and AI management system, namely ISO/IEC 23894[7] and ISO/IEC 42001[8], have been found insufficient in meeting the requirements of the AI Act [44].

---

[7] https://www.iso.org/standard/77304.html.
[8] https://www.iso.org/standard/81230.html.

Indeed, within the current AI standardisation realm, the two main international standards covering AI risk management aspects that have been analysed in regard to AI risk management in the technical report published by the European Commission's Joint Research Centre [44] are: (i) **ISO/IEC 42001 "Information technology - Artificial intelligence - Management system"** that offers a framework to assist with implementation of AI management systems, and (ii) **ISO/IEC 23894 "Artificial intelligence - Guidance on risk management"**, which offers practical guidance on managing risks, albeit defined as the effect of uncertainty, without relation to potential harms to individuals. Due to this and other reasons, the requirements defined in these standards are not useful to comply with the provisions of the AI Act in regard to risk management [44], however they may be useful to identify relevant technical information elements for the development of AI Cards while we await harmonised standards for the AI Act. In this section, we elaborate on information elements for the documentation of an AI risk management system by obtaining insights from these two standards.

We used the overall structure of AI management systems, which follows the ISO/IEC's *harmonised structure* for management system standards[9], as a resource for extracting the key organisational activities within an AI risk management system. For each activity, documentation needs, in terms of information elements, were identified from both ISO/IEC 42001 and 23894 (for a summary of the analysis see Fig. 2). These information elements can be categorised into four overall groups:

- Information about the *context of the AI system and the organisation*, for example, the AI system's intended purpose and the role of the organisation in relation to the system.
- Details of the *risk management system* in place, e.g. the policies, responsibilities, and resources required for implementation of the risk management system itself. This category of information is relevant to the ISO/IEC 23894's AI risk management framework, whose intention is to help with AI governance and integration of AI risk management activities into an organisation's existing processes.
- Documentation of *risk management processes* across different phases: planning (ex-ante), operation (ongoing), and post-operation (ex-post).
- *Results of AI risk management*, which can be represented in artefacts produced throughout the risk management process, e.g. risk assessment documentation that lists AI risks identified, their likelihood, severity, sources, consequences, and impacts.

## 4    AI Cards

Extensiveness of technical documentation and confidentiality of its content may hinder collaboration and communication with AI stakeholders-many of whom

---

[9] https://www.iso.org/sites/directives/current/consolidated/index.html.

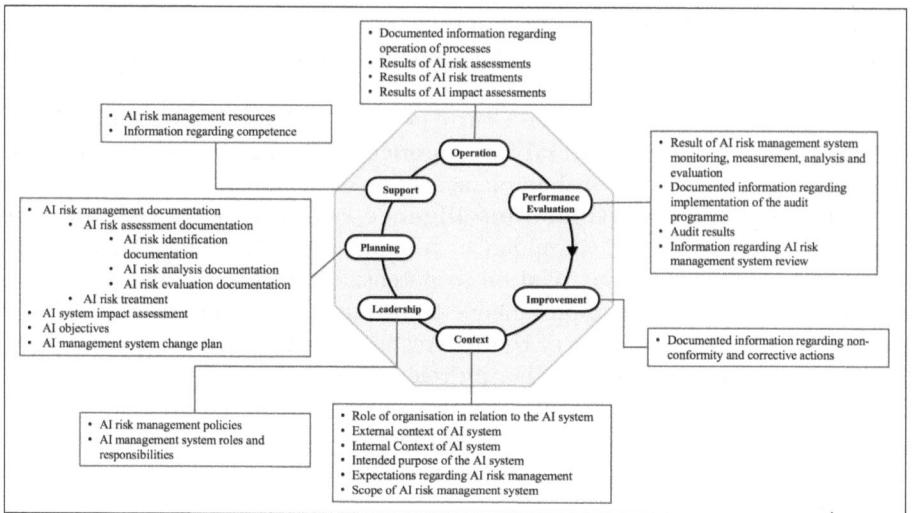

**Fig. 2.** Information elements identified for AI risk management system documentation (note they are inspired by ISO/IEC 42001 and 23894, and therefore will be updated according to the AI Act's future official harmonised standards).

are not necessarily technical AI experts. To address this challenge, we propose the **AI Cards framework**, as a structured information sheet, which offers an overview of a use of an AI system and its related trustworthy AI concerns by inclusion of crucial information about technical specification, context of use, risks, and compliance. With an intuitive representation, the AI Cards framework introduces a simple, transparent, and comprehensible summary card laying out a holistic picture of a specific AI use case and its risks without disclosing sensitive information. The framework encompasses machine-readable specifications that enhance interoperability, enable automation, and support extensibility.

## 4.1   Development Process

In shaping the framework, we first specified its requirements, which are summarised in Table 1. Based on the defined scope, we re-examined our analysis of Annex IV. The information elements within the scope were considered as candidates for inclusion in the AI Cards. However, many of them were too detailed to be represented in a summary card. Therefore, we propose condensed views for these information elements through visual aids.

The AI Cards framework was defined and iteratively refined in consultation with researchers involved in digital policymaking within the European Commission and further validated by conducting an online anonymous survey with law and technology researchers (see Sect. 6 for more details).

**Table 1.** AI Cards framework requirements

| Purpose | Providing an overview of technical documentation that effectively conveys key information |
|---|---|
| Scope | In the scope: AI system, as a whole, its context of use, and information relevant to trustworthy AI concerns including risk management system. Out of the scope: organisational processes, details of management systems, and documentation of AI components, including data and model. The framework is horizontal and does not take sector-specific nuances into account |
| Key audience | AI users, end-users, subjects, providers, developers, auditors, and policymakers |
| Representation formats | An easy-to-understand visual human-readable representation accompanied with a machine-readable specification |

## 4.2    Information Elements

Figure 3 shows a visualisation of the AI Card which condenses the information elements in 9 sections. To enable representing AI Cards in a time series and to allow version control required to reflect the evolving nature of AI systems, the Card's metadata includes essential information describing its **version, issuance date, language, publisher,** and **contact** information. Further, the **URL of the machine-readable specification** is included.

**(1) General Information.** This section provides the key information about the AI system including its **version, modality,** e.g. standalone software or safety component of a product, main **AI techniques** used, **provider(s),** and **developer(s).**

**(2) Intended Use.** The AI Act put a considerable emphasis on intended use of the system, given its profound effect on risks and impacts. For describing the intended use, we propose using the combination of the 5 concepts identified in [18], which are:

- the **domain** in which the system is intended to be used within,
- the **purpose,** i.e. end goal, of using the system in a specific context,
- the **AI capability** that enables realisation of the purpose,
- the **AI deployer** which is the entity using the AI system,
- the **AI subject** which is the entity subjected to the outputs, e.g. decisions, generated by the AI system.

Noteworthy, the subsequent sections heavily rely of the information represented in this section, considering that they should be defined in the view of the intended use.

**(3) Key Components.** Many AI systems are built through integration of multiple AI models, datasets, general-purpose AI systems, and other software elements, each of which has an effect on the system's behaviour and in turn its risks [30]. This section provides the system's *high-level architecture* in terms

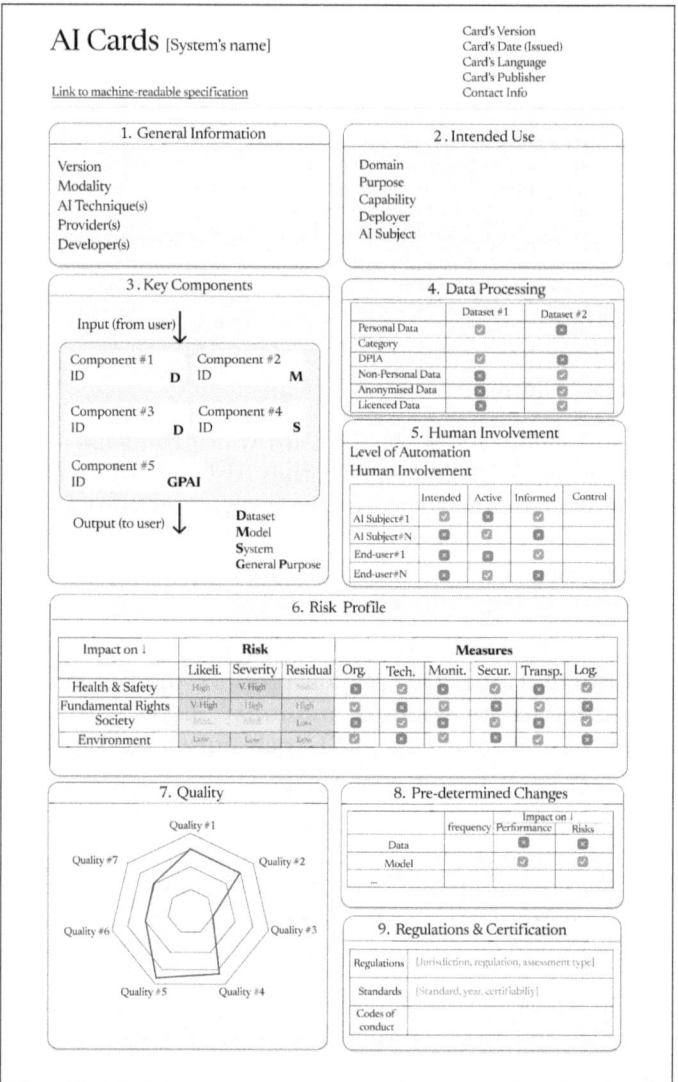

**Fig. 3.** Visual representation of AI Cards.

of incorporating **components**. For each component, its **name**, **version**, and **link to documentation or ID** are presented. For detailed information about components, we rely on their documentation provided presumably by the components' providers in the form of **information sheets**, e.g. Datasheets, Model Cards, and AI Factsheets.

**(4) Data Processing.** AI systems that do not process data, if they exist at all, are rare. Within the EU digital acquis, the GDPR [1], which protects the fundamental right to privacy by regulating *personal data*, is applicable to those AI systems that process natural persons' data. Therefore, having knowledge of whether a given use of an AI system involves processing of personal data is crucial to correctly interpret the resulting legal compliance obligations. This section specifies inclusion of processing **personal data** and shows the **category** of the data, e.g. biometric data, and represents whether data protection impact assessment **(DPIA)** is conducted. It further specifies inclusion of **non-personal**, **anonymised**, and **licenced** data.

**(5) Human Involvement.** Involvement can take different forms depending on the phase of AI development, the role of human actors, and the system's **level of automation**-which has a range from fully autonomous to fully human-controlled according to ISO/IEC 22989[10]. The level of automation also has a substantial effect on the safeguards, including human oversight measures, required for controlling AI risks. This section provides an overview of involvement of two specific human actors: **AI end-users**, who use the AI system's output and **AI subjects**, who are subjected to the outputs of the system. For these actors, we look into the following aspects of involvement:

- **Intended involvement**: represents whether the involvement of a specific actor is as intended. An example of an *intended* AI subject in an AI-based proctoring system is a student sitting an online test. In this case, other persons present in the room are *unintended* AI subjects.
- **Active involvement**: shows whether a specific actor actively interacts with the AI system.
- **Informed involvement**: represents whether a specific actor was informed that an AI system is in place; for example, in cases where a decision affecting a person's education is made using AI-based solutions.
- **Control over AI outputs**: shows the extent to which AI subjects and end-users have control over AI outputs, in particular decisions made by the AI system and their impacts. Inspired by OECD's modes of operationality [35], we consider the following six levels of control:
  - An AI subject/end-user can opt in the system's output.
  - An AI subject/end-user can opt out of the system's output.
  - An AI subject/end-user can challenge the system's output.
  - An AI subject/end-user can correct the system's output.
  - An AI subject/end-user can reverse the system's output ex-post.
  - An AI subject/end-user cannot opt out of the system's output.

**(6) Risk Profile.** This section provides a high-level summary of risk management results, which includes an overview of **likelihood**, **severity** and **residual risk** associated with risks that have impact on areas of **health and safety**,

---

[10] https://www.iso.org/standard/74296.html.

**fundamental rights, society**, and **environment**. Further, this section shows whether any technical, monitoring (human oversight), cybersecurity, transparency, and logging measures applied to control the risks.

**(7) Quality.** With many AI incidents rooted in poor quality, ensuring that the system is of high quality by participating in AI regulatory sandboxes, benchmarking, and performing tests is necessary. We propose illustrating key **AI system qualities** using a radar chart. We recommend including **Accuracy, robustness**, and **cybersecurity**, which are the key qualities explicitly mentioned in the Act (Art. 15). Further, the relevant *product quality* and *quality in use*, described by ISO/IEC 25059 on AI SQuaRE (Systems and software Quality Requirements and Evaluation)[11] can be considered to be included in the quality section.

**(8) Pre-determined Changes.** This section provides a list of **pre-determined changes** to the system and its performance in terms of **subject** and **frequency** of change as well as the potential **impacts of change on performance and risks**.

**(9) Regulations & Certification.** This section lists the main digital **regulations** the AI system is compliant with, key **standards** to which the system or the provider(s) conform, and **codes of conduct** followed in development or use of the AI system.

### 4.3   Machine-Readable Specification

The visual representation assists stakeholders in gaining an understanding of an AI system, its context, and the associated trustworthy AI concerns without delving into the extensive details of technical documentation. However, these are not sufficient to support some of the desirable features of documentation, including search and tracking capabilities, conducting meta-analysis, comparing multiple AI systems, and automating generation and update of documentation [21]. To include these features, the AI Cards framework supports machine-readable representation of information by leveraging the standards, methods, and tools provided by the World Wide Web Consortium (W3C)[12], motivated by the body of work discussed in Sect. 2.3 as well as the rise in uptake of open data formats for documentation, reporting, and sharing information by the AI community, e.g. HuggingFace's use of JSON Model Cards[13], and by the EU, e.g. machine-readable regulatory reporting [15], DCAT-AP open data portals[14], and

---

[11] https://www.iso.org/standard/80655.html.
[12] https://www.w3.org/.
[13] https://huggingface.co/docs/hub/model-cards.
[14] https://op.europa.eu/en/web/eu-vocabularies/dcat-ap.

EU vocabularies and ontologies[15]. The machine-readable representation, which relies on Semantic Web technologies, fosters openness and interoperability-both essential for exchanging information across the AI value chain. This representation is extensible and therefore enables accommodating sector-specific information requirements and allows adaptation to highly-anticipated guidelines from authorities including the AI office, the potential amendments to the Act via delegated Acts, and case law. Grounded on formal logic, it also assists in ensuring that the information is complete, correct, and verifiable.

For provision of machine-readable specifications for AI Cards, an *ontology* is needed to provide a common semantic model that explicitly represents terms, their definitions, and the semantic relations between them, ensuring consistency and interoperability. For this, building upon our previous work [17], we further extend AIRO to support modelling information elements featured within the AI Cards. This extension borrows concepts and relations from Data Quality Vocabulary (DQV)[16] [6] for expressing AI quality and Data Privacy Vocabulary (DPV)[17] [37] for modelling personal data processing. Figure 4 depicts key classes and relations required for representing information featured in each section of the AI Card.

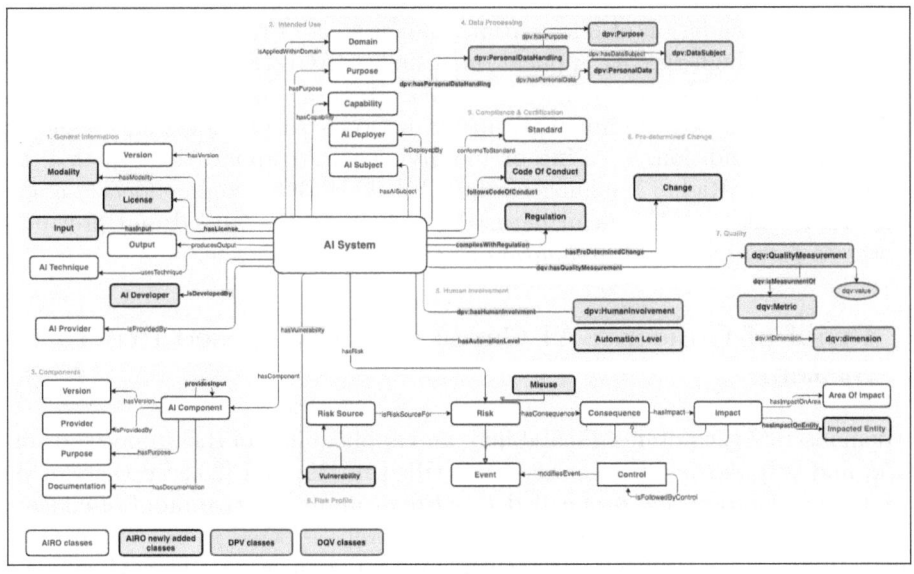

**Fig. 4.** An extension of AIRO for AI Cards.

With the ontology providing the schema, the information incorporated in the AI Cards can be represented as an RDF graph. This opens the door to leveraging

---

[15] https://op.europa.eu/en/web/eu-vocabularies/controlled-vocabularies.

[16] https://www.w3.org/ns/dqv.

[17] https://w3id.org/dpv/.

the power of Semantic Web standards and technologies across a variety of tasks, among them are:

- Querying to retrieve information about an AI system and information related to demonstration and investigation of legal compliance, for example verifying presence of measures to address potential AI impacts on the fundamental right to non-discrimination in support of compliance with Art. 9 of the AI Act. The information retrieval is implemented using SELECT queries in the SPARQL[18], which the W3C standardised query language, as shown in [17].
- Updating information about an AI application, for example when a mitigation measure is no longer effective and is replaced with a new measure, which is implemented using DELETE/INSERT SPARQL queries.
- Integrating information provided in documentation of incorporating components (demonstrated in the "Key Componets" section of the AI Card), for example a Model Card that documents a system's incorporating model, when provided in standardised linked data formats such as JSON-LD or RDF.
- Reasoning that might support conformity assessment and legal compliance tasks, for example defining semantic rules to suggest the regulatory risk profile associated with an AI application. Semantic rule-checking using the Shapes Constraint Language (SHACL)[19], which is a W3C standardised language, is valuable in validation of information against a set of rules (see prior work on determining high-risk AI applications using SHACL [18]).
- Expressing intended use policies in agreements between different parties in the AI value chain, for example agreements between AI providers and deployers describing conditions of using or modifying an AI application. In this, the Open Digital Rights Language (ODRL)[20]-the W3C recommended language for representing policies-can be used for describing intended and precluded uses as *permission* and *prohibition* statements respectively.

## 5   Proof-of-Concept: AI Cards for an AI-Based Proctoring System

To demonstrate the potential scalability, and applicability of the proposed framework, and inspired by the use cases described in [39] and [23], we take an AI-based student proctoring tool called *Proctify* as an illustrative proof-of-concept. It is important to note that this system is merely selected to demonstrate the applicability of the AI Cards framework in real-world cases. The AI Act generally considers such proctoring systems as *high-risk* in the education domain when used for monitoring and detecting suspicious behaviour of students during tests [25].

Proctify is intended to detect suspicious behaviour during online exams by analysing facial behaviour from a student's facial video captured throughout

---

[18] https://www.w3.org/TR/sparql11-query/.

[19] https://www.w3.org/TR/shacl/.

[20] https://www.w3.org/TR/odrl-model/.

the exam using a webcam. Prior to this, students have explicitly consented to be recorded during the exam and informed that they must be alone in the room. The system incorporates a graphic interface displaying an analysis of the student's face including the head pose, gaze direction, and face landmarks' positions. This extracted information is then provided as an input to *SusBehavedModel*, which has been trained in-house by the system's provider using *SusBehavedDataset*, to determine whether the student is displaying suspicious behaviour, e.g. looking away from the screen, leaving the room, or a third person detected in the room. Detection of suspicious behaviour raises an alarm in the interface to inform and let the human oversight actors, e.g. human instructors, take appropriate actions, e.g. communicating with the student.

Throughout the risk management process, the risks and impacts of the system are identified and assessed by the provider, including the following: the system may have lower accuracy for students with darker skin tones and a higher rate of false-positive alarms for students wearing glasses. Further, false-negatives and false-positives are more frequent for students with health issues or disabilities that affect their facial behaviour. There is also a chance of over-reliance of human instructors on the system's output (automation bias). These events have the potential to negatively impact students' *mental health*, *future career*, and their rights to *dignity* and *non-discrimination*. Some of the measures applied to address the system's risks and impacts are: ensuring the dataset is representative and diverse in demographic terms, conducting rigours and frequent testing of accuracy, assigning expert human proctors, and creating clear protocols to act upon when an alarm is raised. The Proctify's AI Card is visualised in Fig. 5 and its machine-readable specification is available online[21]. As shown in the figure, the visual representation of the AI Card provides a summary of risk management information without disclosing the details. However, the machine-readable specification is capable of modelling the details.

## 6    Validation

As mentioned earlier, to identify essential information elements of the AI Cards framework, we consulted experts involved in EU digital policymaking, in particular the AI Act, whose backgrounds are in AI transparency and documentation, explainability, cybersecurity, standardisation, and Semantic Web. Throughout the development process, our analysis of the AI Act's provisions, the visual representation of AI Cards, and its machine-readable specification were discussed to ensure alignment with EU digital policies, adoption of correct terminology, and suitability of the AI Cards framework for addressing common concerns of AI stakeholders. After reaching a solid structure for the AI Cards, an online anonymous survey has been conducted to assess the usefulness of the AI Cards framework.

---

[21] https://github.com/DelaramGlp/airo/blob/main/usecase/proctify.ttl.

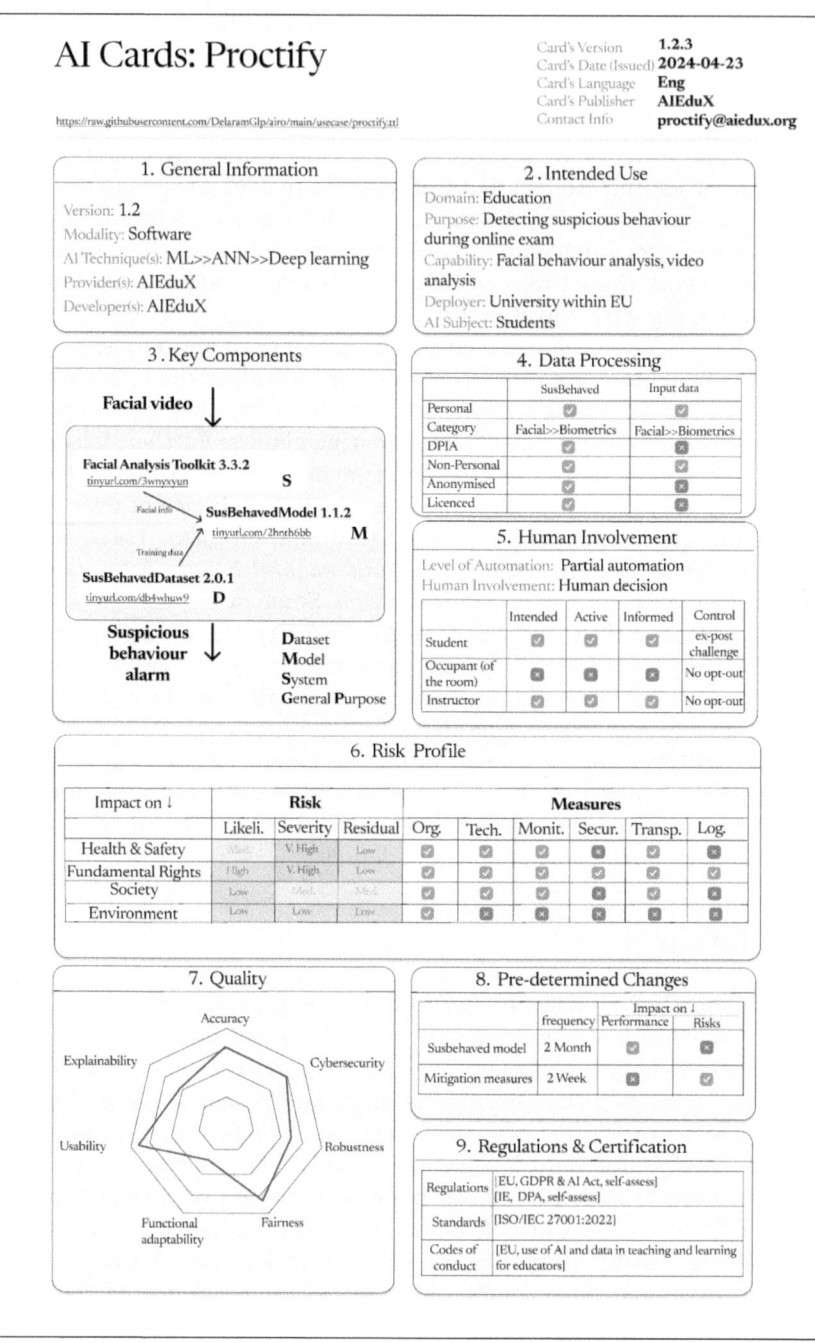

**Fig. 5.** An example of AI Card for an AI-based proctoring system.

**Survey Design.** In addition to questions regarding the respondent's background, the survey includes questions about usefulness of (i) (only) the visual representation, (ii) (only) the machine-readable representation, and (iii) the overall AI Cards framework (both human- and machine-readable representations). In the latter, to assess how stakeholders envisage usability of the AI Card the System Usability Scale (SUS) [11] is used. As the SUS is originally designed for assessing perceived usability, minor wording changes were applied to fit it with the purpose of AI Cards' assessment and to provide more clarity. Table 2 shows the list of SUS-based questions for assessing the AI Cards' usability, each answered on a 5-point Likert scale ranging from 1 (strongly disagree) to 5 (strongly agree).

**Recruitment.** In the first phase, a student cohort of participants $(N = 23)$ were recruited from the Data Governance module in Dublin City University attended by students enrolled in Masters of Art (MA) in Data Protection and Privacy (MDPP) and the European Master in Law, Data and Artificial Intelligence (EMILDAI). It is important to highlight that, because of the nature of this module, some of the participants (30 percent) have notable positions in public organisations, industry, and NGOs. It should be noted that a second phase of evaluation with policymakers and industry stakeholders is currently underway.

**Preliminaries.** Prior to carrying out the survey, it was reviewed and approved by the ethics committee within the school of computer science and statistics of Trinity College Dublin. To start, in a preliminary interactive session, information about the AI Act and the AI Cards was presented to participants and they were asked to sign the informed consent form. Then, they could start completing the survey. It should be noted that the participation in the survey was optional.

**Findings.** 23 responses were collected in this first phase of the evaluation from the student cohort. Regarding the usefulness of the visual representation, inclusion of the following information was suggested in the survey results:

- information related to the registry process,
- link to the GDPR compliance including summary of the requirements for data processing and retention period of data if personal data processed,
- limitations of the system,
- information regarding tests including tests, bugs, issues, and date of testing,
- a legend or key to explain the abbreviations.

In the context of implementation and enforcement of the AI Act, the participants agreed on the usefulness of the human-readable representation in compliance checking, creating technical and risk management system documentation, and creating guidelines. The participants further referred to the following potential uses of the AI Cards: *ensuring traceability and transparency, monitoring the system, reviewing periods for pre-determined changes, raising awareness,* and *explaining the core concepts to stakeholders with different levels of AI literacy and technical knowledge.* In regard to usefulness of AI Cards as a tool to facilitate exchange of information regarding an AI system and its risks, while all participants acknowledged its usefulness to some extent for communication within the AI development ecosystem and with authorities, a minority (4 out of 23) did not perceive it valuable for communication with the public.

Regarding machine-readable specifications, there was general consensus on its usefulness for automated tools, establishing a common language, and structuring the EU registry of AI systems. Further, the participants indicated that the machine-readable specifications could assist in *determining the legal risk category, providing a better understanding of models,* and *building models with better compliance capabilities.*

Regarding the potential target users of the framework, one participant referred to *"AI Brokers for comparing AI Systems - would require more metrics but a good start for comparison"* and another participant mentioned *"Members of the public, in fostering awareness and understanding of AI systems".* The results of SUS usability test are shown in Table 2, where due to the mixed tone of the items a higher score for odd-numbered items and a lower score for even-numbered items are desired. The average SUS score, calculated using the formula proposed in [27], is 66.30.

**Table 2.** SUS questions used for evaluating usability of AI Cards

| No. | Question | Score (mean) |
|-----|----------|--------------|
| 1 | I think that AI stakeholders would like to use the AI Cards framework frequently | 4.21 |
| 2 | I find the AI Cards framework unnecessarily complex | 2.13 |
| 3 | I think the AI Cards framework is easy to use | 3.96 |
| 4 | I think that AI stakeholders would need the support of a technical person to be able to use the AI Cards framework | 3.35 |
| 5 | I find the various aspects (i.e. information elements and human- and machine-readable formats) in the AI Cards framework are well integrated | 4.22 |
| 6 | I think there is too much inconsistency in the AI Cards framework | 1.83 |
| 7 | I would imagine that most people would learn to use the AI Cards framework very quickly | 3.91 |
| 8 | I find the AI Cards framework very cumbersome to use | 2.26 |
| 9 | I feel very confident using the AI Cards framework | 3.43 |
| 10 | I need to learn a lot of things before I could get going with the AI Cards framework | 3.30 |

Another key aspect in evaluation of effectiveness and usefulness of an AI documentation framework is the extent of its adoption by the AI community, as evidenced in widely-adopted documentation approaches, e.g. Datasheets [16],

Model Cards [32], and Factsheets [9]. Given that the AI Act was only adopted recently and the final text has not been published yet, the extent of AI Cards' adoption will need to be assessed over time. As indicated by one the participants, *"The real world application would be interesting to see and may require changing based on different Member State authorities requiring other information be made available for documentation purposes"*.

# 7 Benefits and Potential Applications

The AI Cards framework distinguishes itself from existing documentation approaches through:

- Its **alignment with the EU AI Act's provisions** in regard to technical and risk management system documentation. It should be noted that though the framework is designed based on the requirements of the EU AI Act, its modular and semantic nature makes it scalable to be applied to document a wide range of AI applications, regardless of the jurisdiction they are used in.
- Its dual approach towards information representation, which makes the framework **accessible** to both humans and machines, facilitates communication, promotes **interoperability**, and enables automation.
- Its holistic **AI system-focused** approach, which allows inclusion of information regarding the context of use, risk management, and compliance in addition to technical specifications. While most existing documentation practices address potential risks or ethical issues at a very high and unstructured level [24], an AI Card draws a picture of the **risk management system** that is already established by illustrating an overview of identified risks and applied mitigation measures.
- Its **modular and future-proof** design that enables reconfiguration and extension of the framework to address documentation requirements arising from forthcoming AI regulations, policies, and standards.

AI Cards' machine-readable specification adds an extra level of potential for application. In terms of documentation management, this specification facilitates frequent modification, version control, and integration of data from different sources. Further, it lays the ground for development of supporting tools and RegTech solutions. Within the context of the EU AI Act, these supporting tools can be employed to assist *AI providers* in documentation and exchange of information required for compliance, help *conformity assessment bodies* in tasks related to auditing and certification, and assist *AI deployers* in conducting fundamental rights impact assessment (FRIA). Further, AI Cards might become useful for the *AI Office* in the development of automated Semantic Web-based tools to assist with FRIA, which has an overlap with technical documentation in terms of information requirements. Currently, there is little guidance for how to conduct risk assessment on the impact of AI systems on fundamental rights and the state of the art in taking fundamental rights seriously in the context of AI technical standardisation is new and fast evolving. Therefore, the locating, consumption and reuse of examples of AI risks assessment and accompanying technical documentation is urgently needed to establish a public knowledge base from which AI providers and deployers, especially less well-resourced SMEs and public bodies, can draw legal certainty. Such a sharing of AI risks and the documentation in AI card may therefore accelerate regulatory learning both between regulatory bodies and between prospective AI providers and deployers. Where confidentiality concern may impede the open access sharing of such information, official regulatory learning structures such as sandboxes and testing with human subjects could use this mechanism to maximise sharing and learning between participating stakeholders. Shared searchable repositories of AI risks and corresponding AI Cards may also offer support in identifying risks of reasonably foreseeable misuse of AI systems.

The semantic model, provided as an open-source ontology, can function as a basis for a pan-European AI vocabulary[22] to help establish a common language across different and multidisciplinary actors involved in the AI ecosystem. This consistency in the language, accompanied by a machine-readable representation, promotes interoperability and streamlines the information exchange required for incident reporting, compliance checking, and sharing best practices. It also promotes broader participation of stakeholders needed in development of standards for the AI Act [46]. In addition, it can be utilised for structuring the EU database of high-risk AI systems (Art. 60) and incident reports (Art. 62). As an added advantage, the semantic model can further be improved to evolve into a multilingual ontology supporting official EU languages.

In a broader context, the semantic model is helpful for AI providers and users that operate in different jurisdictions in addressing challenges of cross-border compliance and interoperability by providing an extensible and adaptable structure to maintain information. Provided as an open resource, the schema can be reused and enhanced by the AI community to fit their needs in regard to documentation and risk management.

## 8   Conclusion and Further Work

With this paper, we took an initial step to address the current lack of standardised and interoperable AI documentation practices in alignment with the EU AI Act. We proposed AI Cards, a novel framework for documentation of uses of AI systems in two complementary human- and machine-readable representations. We envision this contribution as a valuable input for future EU policymaking and standardisation efforts related to technical documentation, such as Implementing Acts, European Commission-issued guidelines, and European standards. We also hope that this work promotes use of standardised and interoperable Semantic Web-based formats in AI documentation.

In an ongoing effort to enhance AI Cards' usability and wider applicability, an in-depth user experience study (second phase of the evaluation) involving AI

---

[22] See existing examples of EU vocabularies here: https://op.europa.eu/en/web/eu-vocabularies.

practitioners, auditors, standardisation experts, and policymakers is underway. Our future work also takes the direction towards development of automated tools for generating AI Cards from users' input and existing AI documents provided in textual formats. Further, alignment of AI Cards with documentation and reporting requirements of EU digital regulations, including the GDPR, Digital Services Act (DSA) [5], Interoperability Act [4], and Data Governance Act (DGA) [3], as well as well-known risk management frameworks, e.g. NIST Artificial Intelligence Risk Management Framework (AI RMF 1.0) [34], is a key future step in refining the AI Cards.

**Acknowledgments.** This project has received funding from the European Union's Horizon 2020 research and innovation programme under the Marie Skłodowska-Curie grant agreement No 813497 (PROTECT ITN), as part of the ADAPT SFI Centre for Digital Media Technology is funded by Science Foundation Ireland through the SFI Research Centres Programme and is co-funded under the European Regional Development Fund (ERDF) through Grant#13/RC/2106_P2.

**Disclosure of Interests.** The views expressed in this article are purely those of the authors and should not, under any circumstances, be regarded as an official position of the European Commission.

# References

1. Regulation (EU) 2016/679 of the European parliament and of the council of 27 April 2016 on the protection of natural persons with regard to the processing of personal data and on the free movement of such data, and repealing directive 95/46/EC (general data protection regulation) (2016). http://eur-lex.europa.eu/legal-content/EN/TXT/?uri=OJ:L:2016:119:TOC
2. Proposal for a regulation of the European parliament and of the council laying down harmonised rules on artificial intelligence (artificial intelligence act) and amending certain union legislative acts) and amending certain union legislative acts (2021). https://eur-lex.europa.eu/legal-content/EN/TXT/?uri=CELEX:52021PC0206
3. Regulation (EU) 2022/868 of the European parliament and of the council of 30 may 2022 on European data governance and amending regulation (EU) 2018/1724 (data governance act) (2022)
4. Proposal for a regulation of the European parliament and of the council laying down measures for a high level of public sector interoperability across the union (interoperable Europe act) (2022). https://eur-lex.europa.eu/legal-content/EN/TXT/?uri=CELEX:52022PC0720
5. Regulation (EU) 2022/2065 of the European parliament and of the council of 19 October 2022 on a single market for digital services and amending directive 2000/31/EC (digital services act) (2022)
6. Albertoni, R., Isaac, A.: Introducing the data quality vocabulary (DQV). Semantic Web **12**(1), 81–97 (2021)
7. Amith, M.T., et al.: Toward a standard formal semantic representation of the model card report. BMC Bioinform. **23**(6), 1–18 (2022)
8. Araujo, T., Helberger, N., Kruikemeier, S., De Vreese, C.H.: In AI we trust? perceptions about automated decision-making by artificial intelligence. AI Soc. **35**, 611–623 (2020)

9. Arnold, M., et al.: FactSheets: increasing trust in AI services through supplier's declarations of conformity. IBM J. Res. Dev. **63**(4/5), 6-1 (2019)
10. Balahur, A., et al.: Data quality requirements for inclusive, non-biased and trustworthy AI (2022)
11. Brooke, J.: SUS-a "quick and dirty" usability scale. Usability Eval. Ind. 189–194 (1996)
12. European Commission: Draft standardisation request to the european standardisation organisations in support of safe and trustworthy artificial intelligence (2022). https://ec.europa.eu/docsroom/documents/52376
13. Donald, A., et al.: Towards a semantic approach for linked dataspace, model and data cards. In: Companion Proceedings of the ACM Web Conference 2023, pp. 1468–1473 (2023)
14. European Commission and Directorate-General for Communications Networks, Content and Technology: Ethics guidelines for trustworthy AI. Publications Office (2019). https://doi.org/10.2759/346720. https://data.europa.eu/doi/10.2759/346720
15. European Commission and Directorate-General for Financial Stability, Financial Services and Capital Markets Union: MRER proof of concept – Assessing the feasibility of machine-readable and executable reporting for EMIR. Publications Office of the European Union (2022). https://doi.org/10.2874/036007
16. Gebru, T., et al.: Datasheets for datasets. Commun. ACM **64**(12), 86–92 (2021)
17. Golpayegani, D., Pandit, H.J., Lewis, D.: AIRO: an ontology for representing AI risks based on the proposed EU AI act and ISO risk management standards, pp. 51–65. IOS Press (2022)
18. Golpayegani, D., Pandit, H.J., Lewis, D.: To be high-risk, or not to be-semantic specifications and implications of the AI act's high-risk AI applications and harmonised standards. In: Proceedings of the 2023 ACM Conference on Fairness, Accountability, and Transparency, pp. 905–915 (2023)
19. Gruetzemacher, R., Whittlestone, J.: The transformative potential of artificial intelligence. Futures **135**, 102884 (2022)
20. Gyevnara, B., Fergusona, N., Schaferb, B.: Get your act together: a comparative view on transparency in the AI act and technology. arXiv preprint arXiv:2302.10766 (2023)
21. Heger, A.K., Marquis, L.B., Vorvoreanu, M., Wallach, H., Wortman Vaughan, J.: Understanding machine learning practitioners' data documentation perceptions, needs, challenges, and desiderata. Proc. ACM Hum.-Comput. Interact. **6**(CSCW2), 1–29 (2022)
22. Holland, S., Hosny, A., Newman, S., Joseph, J., Chmielinski, K.: The dataset nutrition label. Data Prot. Privacy **12**(12), 1 (2020)
23. Hupont, I., Fernández-Llorca, D., Baldassarri, S., Gómez, E.: Use case cards: a use case reporting framework inspired by the European AI act. Ethics Inf. Technol. **26**(2) (2024)
24. Hupont, I., Micheli, M., Delipetrev, B., Gómez, E., Garrido, J.S.: Documenting high-risk AI: a European regulatory perspective. Computer **56**(5), 18–27 (2023)
25. Hupont, I., Tolan, S., Gunes, H., Gómez, E.: The landscape of facial processing applications in the context of the european AI act and the development of trustworthy systems. Sci. Rep. **12**(1), 10688 (2022)
26. Hutchinson, B., et al.: Towards accountability for machine learning datasets: practices from software engineering and infrastructure. In: ACM Conference on Fairness, Accountability, and Transparency, pp. 560–575 (2021)

27. Lewis, J.R.: The system usability scale: past, present, and future. Int. J. Hum.-Comput. Interact. **34**(7), 577–590 (2018). https://doi.org/10.1080/10447318.2018.1455307
28. Maragno, G., Tangi, L., Gastaldi, L., Benedetti, M.: Exploring the factors, affordances and constraints outlining the implementation of artificial intelligence in public sector organizations. Int. J. Inf. Manage. **73**, 102686 (2023)
29. Mazzini, G., Scalzo, S.: The proposal for the artificial intelligence act: considerations around some key concepts. La via europea per l'Intelligenza artificiale, Camardi (a cura di) (2023)
30. Micheli, M., Hupont, I., Delipetrev, B., Soler-Garrido, J.: The landscape of data and AI documentation approaches in the European policy context. Ethics Inf. Technol. **25**(4), 56 (2023)
31. Miron, M., Tolan, S., Gómez, E., Castillo, C.: Evaluating causes of algorithmic bias in juvenile criminal recidivism. Artif. Intell. Law **29**(2), 111–147 (2021)
32. Mitchell, M., et al.: Model cards for model reporting. In: Proceedings of the Conference on Fairness, Accountability, and Transparency, pp. 220–229 (2019)
33. Naja, I., Markovic, M., Edwards, P., Pang, W., Cottrill, C., Williams, R.: Using knowledge graphs to unlock practical collection, integration, and audit of AI accountability information. IEEE Access **10**, 74383–74411 (2022)
34. National Institute of Standards and Technology: Artificial intelligence risk management framework (AI RMF 1.0) (2023). https://doi.org/10.6028/NIST.AI.100-1
35. OECD: OECD Framework for Classification of AI Systems: a tool for effective AI policies (2022). https://oecd.ai/en/classification
36. Pandit, H.J.: A semantic specification for data protection impact assessments (DPIA). In: Towards a Knowledge-Aware AI: SEMANTiCS 2022-Proceedings of the 18th International Conference on Semantic Systems, 13–15 September 2022, Vienna, Austria, pp. 36–50. IOS Press (2022)
37. Pandit, H.J., Esteves, B., Krog, G.P., Ryan, P., Golpayegani, D., Flake, J.: Data privacy vocabulary (DPV)–version 2. arXiv preprint arXiv:2404.13426 (2024)
38. Pandit, H.J., Ryan, P., Krog, G.P., Crane, M., Brennan, R.: Towards a semantic specification for GDPR data breach reporting. In: Legal Knowledge and Information Systems, pp. 131–136. IOS Press (2023)
39. Panigutti, C., et al.: The role of explainable AI in the context of the AI act. In: Proceedings of the 2023 ACM Conference on Fairness, Accountability, and Transparency, pp. 1139–1150 (2023)
40. Pistilli, G., Muñoz Ferrandis, C., Jernite, Y., Mitchell, M.: Stronger together: on the articulation of ethical charters, legal tools, and technical documentation in ml. In: Proceedings of the 2023 ACM Conference on Fairness, Accountability, and Transparency, pp. 343–354 (2023)
41. Roman, A.C., et al.: Open datasheets: machine-readable documentation for open datasets and responsible AI assessments. arXiv preprint arXiv:2312.06153 (2023)
42. Ryan, P., Brennan, R., Pandit, H.J.: DPCat: specification for an interoperable and machine-readable data processing catalogue based on GDPR. Information **13**(5), 244 (2022)
43. Schuett, J.: Risk management in the artificial intelligence act. Eur. J. Risk Regul. 1–19 (2023)
44. Soler Garrido, J., et al.: Analysis of the preliminary AI standardisation work plan in support of the AI act. Technical report, Joint Research Centre (Seville site) (2023)

45. Tartaro, A.: Towards European standards supporting the AI act: alignment challenges on the path to trustworthy AI. In: Proceedings of the AISB Convention, pp. 98–106 (2023)
46. Veale, M., Zuiderveen Borgesius, F.: Demystifying the draft EU artificial intelligence act-analysing the good, the bad, and the unclear elements of the proposed approach. Comput. Law Rev. Int. **22**(4), 97–112 (2021)
47. Zhang, Y., Wu, M., Tian, G.Y., Zhang, G., Lu, J.: Ethics and privacy of artificial intelligence: understandings from bibliometrics. Knowl.-Based Syst. **222**, 106994 (2021)

# Evaluating Differential Privacy on Correlated Datasets Using Pointwise Maximal Leakage

Sara Saeidian$^{(\boxtimes)}$ ⓘ, Tobias J. Oechtering ⓘ, and Mikael Skoglund ⓘ

KTH Royal Institute of Technology, 100 44 Stockholm, Sweden
{saeidian,oech,skoglund}@kth.se

**Abstract.** Data-driven advancements significantly contribute to societal progress, yet they also pose substantial risks to privacy. In this landscape, *differential privacy* (DP) has become a cornerstone in privacy preservation efforts. However, the adequacy of DP in scenarios involving correlated datasets has sometimes been questioned and multiple studies have hinted at potential vulnerabilities. In this work, we delve into the nuances of applying DP to correlated datasets by leveraging the concept of *pointwise maximal leakage* (PML) for a quantitative assessment of information leakage. Our investigation reveals that DP's guarantees can be arbitrarily weak for correlated databases when assessed through the lens of PML. More precisely, we prove the existence of a pure DP mechanism with PML levels arbitrarily close to that of a mechanism which releases individual entries from a database without any perturbation. By shedding light on the limitations of DP on correlated datasets, our work aims to foster a deeper understanding of subtle privacy risks and highlight the need for the development of more effective privacy-preserving mechanisms tailored to diverse scenarios.

**Keywords:** Pointwise maximal leakage · Differential privacy · Correlated data

## 1 Introduction

In today's data-driven landscape, private and public organizations increasingly rely on data collected from individuals for decision-making and to enhance service provision. While insights obtained from data undoubtedly offer value to societies, it is essential not to overlook the risks to *privacy* that come along with it. As a result, extensive research over the past decades has led to the emergence of various privacy definitions and frameworks.

Among the various definitions proposed, *differential privacy* (DP) [8,10] stands out as the most widely accepted framework for understanding and enforcing privacy. DP has been adopted by both public agencies, such as the U.S. Census Bureau [1], and major industry players like Apple [28], Google [15], and Microsoft [6]. DP assumes that data collected from individuals is stored in a

M. Jensen et al. (Eds.): APF 2024, LNCS 14831, pp. 73–86, 2024.
https://doi.org/10.1007/978-3-031-68024-3_4

database that returns answers to queries in a privacy-preserving manner. Its objective is to reveal population-level insights about the data while preserving the privacy of each individual. Specifically, DP ensures that two databases differing in a single entry, presumably information pertaining to a single individual, cannot be distinguished based on their corresponding query responses. This approach aligns with the fundamental idea that "nothing should be learnable about an individual participating in a database that could not be learned without participation" [11].

Despite its widespread success, several studies have raised concerns that DP may not provide sufficient protection for databases containing *correlated* data [5, 16, 18, 21, 22, 30, 31]. Informally, this is because there may be no one-to-one mapping between individuals and entries in the database, and each person's information may contribute to multiple entries. To illustrate this issue, consider the following example from [18].

*Example 1.* Suppose Bob is part of a medical database where his sensitive attribute can take one of the values $1, \ldots, k$. Assume the database is sampled from a distribution such that when Bob's sensitive attribute is $j$, there are $j \times 10,000$ cancer patients in the data. Suppose an adversary queries the database about the number of cancer patients. Let $\mathrm{Lap}(b)$ denote the zero mean Laplace distribution with scale parameter $b > 0$ (i.e., variance $2b^2$). To answer this query while satisfying 0.1-DP, the mechanism returning the response adds $\mathrm{Lap}(10)$ noise to the true count and releases the result (see Definitions 1 and 2 in Sect. 2). However, in this case, the attacker can infer Bob's sensitive attribute with high probability by dividing the noisy answer by 10,000 and rounding to the nearest integer $j$.

Example 1 and similar ones underscore the necessity for privacy definitions that take into account the data-generating distribution. Consequently, several privacy frameworks have emerged to address this concern, including Pufferfish privacy [20], membership privacy [21], Bayesian differential privacy [30], and coupled-worlds privacy [3], to give a few examples. Among these distribution-dependent frameworks, one that is particularly promising is based on a recent notion of information leakage called *pointwise maximal leakage* (PML) [25, 26].

At its core, PML is an information measure that quantifies the amount of information leaking about a secret to a publicly available and correlated quantity. What sets PML apart is its strong operational meaning in the context of privacy. Specifically, PML is derived by assessing risks posed by adversaries in highly general threat models (see Sect. 2 for details). Moreover, PML exhibits remarkable robustness by considering a wide range of adversaries, and flexibility in its application to various data types. Another noteworthy aspect of PML is that its guarantees and privacy parameters are easily interpretable. Informally speaking, it was shown in [27] that enforcing privacy according to PML aligns with the fundamental principle that "nothing should be learnable about the secret that could not be learned from its distribution alone." Therefore, on an abstract level, the objectives of PML parallel those of differential privacy since PML aims to reveal population-level insights about the data while concealing its intricate details. Furthermore, PML is suitable for guaranteeing privacy in complex data-processing systems through various inequalities that it satisfies, including pre-processing, post-processing, and composition inequalities, among others [25, Lemma 1].

Interestingly, while PML is not a generalization or relaxation of differential privacy, connections have been established between the two frameworks. More precisely, it was shown in [27, Thm. 4.2] that when our goal is to protect a database containing independent entries, then differential privacy is equivalent to restricting the PML of each entry in the database across all possible outcomes of a mechanism. This result provides deep insights for protecting databases containing independent entries. However, it also prompts the question: *What is the relationship between PML and differential privacy in scenarios where the entries in a database are correlated?*

### 1.1 Contributions and Outline

In this work, we establish that in scenarios where the entries in a database are correlated, the PML guarantees of mechanisms satisfying pure DP can be arbitrarily weak. Specifically, we prove that there exists a pure DP mechanism with PML levels arbitrarily close to that of a mechanism which releases individual entries from a database without any perturbation. The significance of this result lies in its quantitative nature. In particular, we rely on analytical arguments to demonstrate that DP may not be a suitable framework for protecting correlated databases, in contrast to its performance in scenarios involving independent entries. Our analysis based on PML also distinguishes our work from previous studies, which often rely on intuition, qualitative examples [19,30,31], or experimental evidence [5,22] to illustrate the weakness of DP in the presence of correlations. Overall, our discussions aim to raise awareness in order to prevent creating a false sense of security that results from applying privacy mechanisms indiscriminately.

The remainder of this paper is organized as follows: Sect. 2 presents preliminaries, including definitions of pure differential privacy and pointwise maximal leakage, highlighting their connections and distinctions. In Sect. 3, we demonstrate that in scenarios involving correlated databases, mechanisms satisfying pure differential privacy can be weak when evaluated through the lens of PML. Section 4 contains a brief discussion about our result and some concluding remarks.

## 2 Preliminaries

### 2.1 Notation

We use uppercase letters to denote random variables, lower case letters to denote realizations, and calligraphic letters to denote sets. In particular, we reserve $X$ for representing some data containing sensitive information, e.g., a dataset containing information about individuals. For simplicity, we assume that the alphabet of $X$, denoted by $\mathcal{X}$, is a finite set. With a slight abuse of notation, we use $P_X$ to describe both a distribution for $X$ and its corresponding probability mass function. Moreover, $\mathcal{P}_\mathcal{X}$ denotes the set of all distributions with full support on $\mathcal{X}$.

Let $Y$ be a random variable representing some information released about $X$, taking values on a set $\mathcal{Y}$. The random variable $Y$ is induced by a *mechanism*, i.e., a conditional probability distribution $P_{Y|X}$. Essentially, $Y$ represents the answer to a query posed about $X$. The set $\mathcal{Y}$ can be finite (e.g., when $P_{Y|X}$ is the randomized response mechanism [29]) or infinite (e.g., when $P_{Y|X}$ is the Laplace mechanism [10]). With a slight abuse of notation, we use $P_{Y|X}$ to denote both the conditional distribution of $Y$ given $X$ as well as its density with respect to a suitable dominating measure, e.g., the counting measure or the Lebesgue measure.

Let $P_{XY}$ denote the joint distribution of $X$ and $Y$. We write $P_{XY} = P_{Y|X} \times P_X$ to imply that $P_{XY}(x,y) = P_{Y|X=x}(y)P_X(x)$ for all $(x,y) \in \mathcal{X} \times \mathcal{Y}$. Furthermore, we write $P_Y = P_{Y|X} \circ P_X$ to represent marginalization over $X$, i.e., to imply that $P_Y(y) = \sum_{x \in \mathcal{X}} P_{Y|X=x}(y)P_X(x)$ for all $y \in \mathcal{Y}$.

We say that the Markov chain $U - X - Y$ holds if random variables $U$ and $Y$ are conditionally independent given $X$, that is, if $P_{UY|X} = P_{U|X} \times P_{Y|X}$. The Markov chain $U - X - Y$ implies that $Y$ depends on $U$ only through $X$ and vice versa. We may think of a $U$ satisfying the Markov chain $U - X - Y$ as either a feature of $X$ or a (randomized) function of $X$.

Finally, $[n] := \{1, \ldots, n\}$ describes the set of all positive integers smaller than or equal to $n$, and $\log(\cdot)$ denotes the natural logarithm.

## 2.2  Differential Privacy

Often called the gold standard of privacy, differential privacy (DP) [10,12] stands as the most widely adopted privacy framework, both in theoretical developments as well as real-world deployments. Conceptually, DP considers scenarios where individuals' data is aggregated into a database, with the aim of responding to queries posed to the database without compromising the privacy of any individual contributor. To achieve this goal, DP guarantees that two databases that differ in only one entry (referred to as "neighboring" databases) produce query responses that are hard to distinguish.

Over nearly two decades of extensive research has led to the development of various DP variants and adaptations, such as approximate DP [9], concentrated DP [4,13], Rényi DP [23], and Gaussian DP [7], among others. However, in this paper, our focus is on the original, and arguably the strongest, form of DP, known as "pure" DP [10].

Let $X = (D_1, \ldots, D_n)$ be a database containing $n$ entries. Given $i \in [n]$, $D_i$ represents the $i$-th entry, which takes values on a finite set $\mathcal{D}$ and $D_{-i} = (D_1, \ldots, D_{i-1}, D_{i+1}, \ldots, D_n)$ represents the database with its $i$-th entry removed. Let $P_X = P_{D_1,\ldots,D_n}$ be the distribution according to which databases are drawn from $\mathcal{X} = \mathcal{D}^n$. To obtain the distribution of the $i$-th entry, we marginalize over the remaining $n-1$ entries, that is, for each $d_i \in \mathcal{D}$ and $i \in [n]$ we have

$$
\begin{aligned}
P_{D_i}(d_i) &= \sum_{d_{-i} \in \mathcal{D}^{n-1}} P_X(d_i, d_{-i}) \\
&= \sum_{d_{-i} \in \mathcal{D}^{n-1}} P_{D_i | D_{-i} = d_{-i}}(d_i) \, P_{D_{-i}}(d_{-i}),
\end{aligned}
\tag{1}
$$

where $d_{-i} := (d_1, \ldots, d_{i-1}, d_{i+1}, \ldots, d_n) \in \mathcal{D}^{n-1}$ is a tuple describing the database with its $i$-th entry removed. Note that this setup is very general since we make no independence assumptions and the entries can be arbitrarily correlated.

Suppose an analyst poses a query to the database, with the answer returned by the mechanism $P_{Y|X}$.

**Definition 1. (Differential privacy).** *Given $\varepsilon \geq 0$, the mechanism $P_{Y|X}$ satisfies $\varepsilon$-differential privacy if*

$$
\max_{\substack{d_i, d_i' \in \mathcal{D}: \\ i \in [n]}} \max_{d_{-i} \in \mathcal{D}^{n-1}} \log \frac{P_{Y|D_i = d_i, D_{-i} = d_{-i}}(\mathcal{E})}{P_{Y|D_i = d_i', D_{-i} = d_{-i}}(\mathcal{E})} \leq \varepsilon,
$$

*for all measurable events $\mathcal{E} \subseteq \mathcal{Y}$.*

Importantly, Definition 1 is agnostic to the distribution of the database, $P_X$. Therefore, differential privacy is a property of the mechanism $P_{Y|X}$ alone. This observation has led to the widespread belief that DP offers strong privacy guarantees for individuals in a database irrespective of the distribution $P_X$. However, in Sect. 3 we will present a quantitative example to demonstrate that a differentially private mechanism can in fact leak a large amount of information about individual entries in scenarios where the entries are correlated.

Next, let us recall one of the most commonly employed differentially private mechanisms, namely the *Laplace mechanism* [10]. The Laplace mechanism is often used to answer numerical queries with bounded $\ell_1$-*sensitivity*. Let $\text{Lap}(\mu, b)$ denote the Laplace distribution with mean $\mu \in \mathbb{R}$ and scale parameter $b > 0$ (i.e., variance $2b^2$).

**Definition 2. (Laplace mechanism).** *Let $f : \mathcal{X} \to \mathbb{R}$ be a query with $\ell_1$-sensitivity*

$$
\Delta_1(f) := \sup_{x_1, x_2 \in \mathcal{X} : x_1 \sim x_2} |f(x_1) - f(x_2)|.
$$

*Given $\varepsilon > 0$ the Laplace mechanism returns a query response according to the distribution $Y \mid X = x \sim \text{Lap}\left(f(x), \frac{\Delta_1(f)}{\varepsilon}\right)$, where $x \in \mathcal{X}$.*

It has been shown in [10] that the Laplace mechanism satisfies $\varepsilon$-DP.

The $\ell_1$-sensitivity of a query $f$ describes the largest change in $f(x)$ upon altering the value of one entry in database $x$. The Laplace mechanism then computes $f$ and perturbs it with Laplace noise scaled according to $\Delta_1(f)$ and $\varepsilon$. Examples of queries that can be answered via the Laplace mechanism include *counting queries*, that is, queries of the form "How many entries in the database satisfy property $A$?," and *histogram queries* [12].

### 2.3    Pointwise Maximal Leakage

Pointwise maximal leakage (PML) [25,26] is a recently introduced privacy measure that enjoys a strong operational meaning and robustness. It measures the amount of information leaking about a secret $X$ to the outcomes of a mechanism $P_{Y|X}$. PML is obtained by evaluating the risk posed by adversaries in two highly versatile threat models: the *randomized functions* model [17] and the *gain function* model of leakage [2].

Below, we define PML using both models.

**Definition 3. (Randomized function view of PML [25, Def. 1]).** *Suppose $X$ is a random variable on the finite set $\mathcal{X}$, and $Y$ is a random variable on a set $\mathcal{Y}$ induced by a mechanism $P_{Y|X}$. According to the randomized function view, the pointwise maximal leakage from $X$ to $y \in \mathcal{Y}$ is defined as*

$$\ell(X \to y) := \log \sup_{U:U-X-Y} \frac{\sup_{P_{\hat{U}|Y}} \mathbb{P}\left(U = \hat{U} \mid Y = y\right)}{\max_{u \in \mathcal{U}} P_U(u)}, \tag{2}$$

*where $U$ and $\hat{U}$ are random variables on a finite set $\mathcal{U}$, $P_U = P_{U|X} \circ P_X$, and the Markov chain $U - X - Y - \hat{U}$ holds.*

Definition 3 can be understood as follows. Let $U$ be a randomized function of $X$. For instance, when $X$ is a database, $U$ can represent a single entry or a subset of the entries in $X$. To quantify the amount of information leakage associated with a single released outcome of the mechanism, denoted by $y$, in (2) we compare the probability of correctly guessing the value of $U$ after observing $y$ in the numerator of (2) with the *a priori* probability of correctly guessing the value of $U$ in the denominator. The probability of correctly guessing $U$ after observing $y$ is assessed by assuming that the adversary uses the best guessing kernel $P_{\hat{U}|Y}$, represented by the supremum over $P_{\hat{U}|Y}$ in the numerator of (2). Similarly, the prior probability of correctly guessing $U$ is $\max_{u \in \mathcal{U}} P_U(u)$. Crucially, Definition 3 results in a highly robust measure of privacy since the posterior-to-prior ratio is maximized over all possible randomized functions of $X$, represented by the supremum over all $U$'s satisfying the Markov chain $U - X - Y$. This makes PML particularly useful when we do not know what feature of $X$ an adversary is interested to guess, or different adversaries may be interested in different features of $X$.

Next, we define PML using the gain function model.

**Definition 4. (Gain function view of PML [25, Cor. 1]).** *Suppose $X$ is a random variable on the finite set $\mathcal{X}$, and $Y$ is a random variable on a set $\mathcal{Y}$*

*induced by a mechanism $P_{Y|X}$. According to the gain function view, the pointwise maximal leakage from $X$ to $y \in \mathcal{Y}$ is defined as*

$$\ell(X \to y) := \log \sup_{g} \frac{\sup_{P_{W|Y}} \mathbb{E}\left[g(X, W) \mid Y = y\right]}{\max_{w \in \mathcal{W}} \mathbb{E}\left[g(X, w)\right]}, \tag{3}$$

*where the supremum is over all non-negative gain functions $g : \mathcal{X} \times \mathcal{W} \to \mathbb{R}_+$ with a finite range, and $\mathcal{W}$ is a finite set.*

Definition 4 can be understood as follows. Consider an adversary whose objective is to construct a guess of $X$, denoted by $W$, in order to maximize the expected value of a non-negative gain function $g$. The gain function $g$ captures the adversary's objective and can be tailored to model a wide array of privacy attacks. For example, when $X$ is a database, $g$ can model *membership inference* attacks or *reconstruction* attacks [14] (see [25] for concrete examples of gain functions). To quantify the amount of information leakage associated with a single released outcome $y$, we compare the expected value of $g$ after observing $y$ in the numerator of (3) with the prior expected value of $g$ in the denominator. The posterior expected gain is assessed using the best kernel $P_{W|Y}$, represented by the supremum over $P_{W|Y}$ in the numerator of (3). Similarly, the prior expected gain is $\max_{w \in \mathcal{W}} \mathbb{E}\left[g(X, w)\right]$. Then, to obtain a privacy measure robust to different types of attacks, the posterior-to-prior ratio of the expected gain is maximized over all possible non-negative $g$'s with a finite range.

While Definitions 3 and 4 offer different approaches to defining PML, we showed in [25, Thm. 2] that, in fact, they are mathematically equivalent. Moreover, both definitions can be simplified to the following concise expression.

**Theorem 1.** (*[25, Thm. 1]*). *Let $P_{XY}$ be a distribution on the set $\mathcal{X} \times \mathcal{Y}$ with the marginal distribution $P_X \in \mathcal{P}_\mathcal{X}$ for $X$. The pointwise maximal leakage from $X$ to $y \in \mathcal{Y}$ is*[1]

$$\ell(X \to y) = D_\infty(P_{X|Y=y} \| P_X), \tag{4}$$

*where $P_{X|Y=y}$ denotes the posterior distribution of $X$ given $y \in \mathcal{Y}$, and*

$$D_\infty(P_{X|Y=y} \| P_X) = \log \max_{x \in \mathcal{X}} \frac{P_{X|Y=y}(x)}{P_X(x)}$$

$$= \log \max_{x \in \mathcal{X}} \frac{P_{Y|X=x}(y)}{P_Y(y)},$$

*denotes the Rényi divergence of order infinity [24] of $P_{X|Y=y}$ from $P_X$.*

In addition to its strong operational meaning and robustness, PML satisfies several useful properties that render it suitable for deployment in complex data-processing systems. Notably, PML satisfies a pre-processing inequality, a post-processing inequality, and increases (at most) linearly under composition

---

[1] We use the convention that $P_{X|Y=y} = P_X$ if $P_Y(y) = 0$. That is, conditioning on outcomes with density zero equals no conditioning.

[25, Lemma 1]. Furthermore, as evident from (4), PML is non-negative and satisfies the bound

$$\ell(X \to y) \leq \log \frac{1}{\min_{x \in \mathcal{X}} P_X(x)}, \tag{5}$$

for all mechanisms and all $y \in \mathcal{Y}$. The right hand side of the above inequality essentially describes the maximum amount of information that can leak about $X$ through any mechanism. In other words, it represents the largest PML across all outcomes of a mechanism that releases $X$ without perturbing it.

## 2.4  Differential Privacy as a PML Constraint

In general, PML and DP offer fundamentally distinct approaches to privacy and differ in several key aspects. PML quantifies the amount of information leaked to an outcome of a privacy mechanism and the secret $X$ may encompass various types of sensitive data. For example, $X$ can be a password, an individual's medical records, or an entire database. In contrast, DP was specifically formulated to protect private databases. More importantly, PML depends on both the mechanism $P_{Y|X}$ and the data-generating distribution $P_X$. Consequently, a mechanism that leaks little information leakage under one distribution may leak a lot of information under another distribution. Conversely, DP depends only on the mechanism.

Despite their differences, in [27, Thm. 4.2], we established that when $X$ is a database containing independent entries, DP is equivalent to restricting the amount of information leaked about each entry across all outcomes of a mechanism. Let $\mathcal{X} = \mathcal{D}^n$ denote the set of all possible databases, and $\mathcal{Q}_{\mathcal{X}}$ denote the set of product distributions in $\mathcal{P}_{\mathcal{X}}$ defined as $\mathcal{Q}_{\mathcal{X}} := \{P_X \in \mathcal{P}_{\mathcal{X}} : P_X = \prod_{i=1}^n P_{D_i}\}$.

**Theorem 2.** (*[27, Thm. 4.2]*). *Given $\varepsilon \geq 0$, the mechanism $P_{Y|X}$ satisfies $\varepsilon$-differential privacy if and only if*

$$\sup_{P_X \in \mathcal{Q}_{\mathcal{X}}} \sup_{y \in \mathcal{Y}} \max_{i \in [n]} \ell(D_i \to y) \leq \varepsilon.$$

Theorem 2 demonstrates that when the entries in a database are independent, DP is adequate for ensuring privacy in the sense of PML. However, it also raises the possibility that when the entries are correlated, DP might fall short in terms of PML. We delve into this topic in the next section.

# 3  Privacy for Correlated Databases: PML Vs. DP

As discussed in Sect. 1, the objective of our work is to understand the relation between PML and DP when $X$ is a database with correlated entries. It turns out that pure DP mechanisms can have poor privacy performance when assessed through the lens of PML. Before we formally state our result, we need to define the PML of a mechanism that releases an entry from the database without perturbation.

Let $X = (D_1, \ldots, D_n)$ be a database. Given a distribution $P_X \in \mathcal{P}_X$ over $X$ and $i \in [n]$, let $P_{D_i}$ denote the distribution of the $i$-th entry in a database, obtained by marginalizing $P_X$ over $P_{D_{-i}}$, described by (1). We use

$$\varepsilon_{\max}(D_i) := \log \frac{1}{\min\limits_{d \in \mathcal{D}} P_{D_i}(d)},$$

to denote the largest amount of information that can leak about $D_i$ through any mechanism. By (5), $\varepsilon_{\max}(D_i)$ is equal to the PML of a mechanism that releases $D_i$ with no randomization.

**Theorem 3.** *For each $\delta > 0$ and $\varepsilon > 0$ there exists a database $X = (D_1, \ldots, D_n)$, $i \in [n]$, a mechanism $P_{Y|X}$ satisfying $\varepsilon$-DP, and $y \in \mathbb{R}$ such that*

$$\ell(D_i \to y) > \varepsilon_{\max}(D_i) - \delta.$$

*Proof.* To prove the statement, we construct a binary database where the entries are strongly correlated: If one entry is zero (resp. one), then the other entries are also likely to be zero (resp. one). We then demonstrate that if we use the Laplace mechanism to answer the counting query on this database, then the resulting PML is significantly high.

Let $X = (D_1, \ldots, D_{n+1})$ be a database containing $n + 1$ binary entries.[2] Suppose $P_{D_1}(0) = 1 - P_{D_1}(1) = \alpha$ with $0 < \alpha < 0.5$. Let $D_- = (D_2, \ldots, D_{n+1})$. Fix a constant $0 < \eta < 1$ and suppose the distribution of $D_-$ depends on $D_1$ as follows:

$$P_{D_-|D_1=1}(d_-) = \begin{cases} \eta, & \text{if } d_- = 1^n, \\ \frac{1-\eta}{2^n - 1}, & \text{otherwise.} \end{cases}$$

$$P_{D_-|D_1=0}(d_-) = \begin{cases} \eta, & \text{if } d_- = 0^n, \\ \frac{1-\eta}{2^n - 1}, & \text{otherwise.} \end{cases}$$

Suppose our goal is to release the empirical frequency of the ones in the database using the Laplace mechanism, i.e., $Y \mid X = x \sim \text{Lap}(\frac{\|x\|_1}{n+1}, b)$, where $\|x\|_1$ denotes the $\ell_1$-norm of $x \in \{0,1\}^{n+1}$. Note that the empirical frequency has global sensitivity $\frac{1}{n+1}$, thus the Laplace mechanism with scale parameter $b = \frac{1}{\varepsilon(n+1)}$ satisfies $\varepsilon$-DP. However, here we show that the Laplace mechanism is insufficient for protecting $D_1$. To demonstrate this, we calculate the PML $\ell(D_1 \to y)$ with $y \leq 0$, which depends on the distributions $P_{Y|D_1=1}, P_{Y|D_1=0}$ and $P_Y$.

First, we calculate $P_{Y|D_1=1}(y)$ assuming $y \leq 0$:

---

[2] We consider a database of size $n + 1$ instead of $n$ for notational convenience.

$$P_{Y|D_1=1}(y) = \sum_{d_-\in\{0,1\}^n} P_{Y|D_1=1,D_-=d_-}(y) \cdot P_{D_-|D_1=1}(d_-)$$

$$= \eta\, P_{Y|D_1=1,D_-=1^n}(y) + \frac{1-\eta}{2^n-1}\sum_{d_-\in\{0,1\}^n\backslash 1^n} P_{Y|D_1=1,D_-=d_-}(y)$$

$$= \frac{\eta}{2b}\exp\left(-\frac{|y-1|}{b}\right) + \frac{1-\eta}{2b(2^n-1)}\sum_{d_-\in\{0,1\}^n\backslash 1^n} \exp\left(-\frac{|y-\frac{1}{n+1}-\frac{\|d_-\|_1}{n+1}|}{b}\right)$$

$$= \frac{\eta}{2b}\exp\left(-\frac{|y-1|}{b}\right) + \frac{1-\eta}{2b(2^n-1)}\cdot\sum_{i=0}^{n-1}\binom{n}{i}\exp\left(-\frac{|y-\frac{1}{n+1}-\frac{i}{n+1}|}{b}\right)$$

$$= \frac{\eta}{2b}\exp\left(-\frac{|y-1|}{b}\right) +$$

$$\frac{1-\eta}{2b(2^n-1)}\left[\sum_{i=0}^{n}\binom{n}{i}\exp\left(-\frac{|y-\frac{1}{n+1}-\frac{i}{n+1}|}{b}\right) - \exp\left(-\frac{|y-1|}{b}\right)\right]$$

$$= \frac{2^n\eta-1}{2b(2^n-1)}\exp\left(\frac{y-1}{b}\right) +$$

$$\frac{1-\eta}{2b(2^n-1)}\exp\left(\frac{y-\frac{1}{n+1}}{b}\right)\cdot\left(1+\exp(-\frac{1}{b(n+1)})\right)^n$$

$$= \frac{1}{2b(2^n-1)}\exp(\frac{y}{b})\left[(2^n\eta-1)\exp\left(-\frac{1}{b}\right) +\right.$$

$$\left.(1-\eta)\exp\left(-\frac{1}{b(n+1)}\right)\left(1+\exp(-\frac{1}{b(n+1)})\right)^n\right].$$

Similarly, we can calculate $P_{Y|D_1=0}(y)$ assuming $y \le 0$:

$$P_{Y|D_1=0}(y) = \sum_{d_-\in\{0,1\}^n} P_{Y|D_1=0,D_-=d_-}(y) \cdot P_{D_-|D_1=0}(d_-)$$

$$= \eta\, P_{Y|D_1=0,D_-=0^n}(y) + \frac{1-\eta}{2^n-1}\sum_{d_-\in\{0,1\}^n\backslash 0^n} P_{Y|D_1=0,D_-=d_-}(y)$$

$$= \frac{\eta}{2b}\exp\left(\frac{y}{b}\right) + \frac{1-\eta}{2b(2^n-1)}\sum_{d_-\in\{0,1\}^n\backslash 0^n} \exp\left(\frac{y-\frac{\|d_-\|_1}{n+1}}{b}\right)$$

$$= \frac{1}{2b}\exp\left(\frac{y}{b}\right)\left[\eta + \frac{1-\eta}{2^n-1}\sum_{i=1}^{n}\binom{n}{i}\exp\left(-\frac{i}{b(n+1)}\right)\right]$$

$$= \frac{1}{2b}\exp\left(\frac{y}{b}\right)\left[\eta - \frac{1-\eta}{2^n-1} + \frac{1-\eta}{2^n-1}\sum_{i=0}^{n}\binom{n}{i}\exp\left(-\frac{i}{b(n+1)}\right)\right]$$

$$= \frac{1}{2b(2^n-1)}\exp\left(\frac{y}{b}\right)\left[2^n\eta-1 + (1-\eta)\left(1+\exp(-\frac{1}{b(n+1)})\right)^n\right].$$

Since $\exp(-x) \leq 1$ for $x \geq 0$, then $P_{Y|D_1=1}(y) \leq P_{Y|D_1=0}(y)$ when $y \leq 0$. Next, we calculate $P_Y(y)$ for $y \leq 0$:

$$P_Y(y) = (1-\alpha)P_{Y|D_1=1}(y) + \alpha P_{Y|D_1=0}(y)$$

$$= \frac{1}{2b(2^n - 1)} \exp\left(\frac{y}{b}\right) \left[\left(2^n \eta - 1\right)(1-\alpha)\exp\left(-\frac{1}{b}\right) + \right.$$

$$(1-\eta)(1-\alpha)\exp\left(-\frac{1}{b(n+1)}\right)\left(1+\exp(-\frac{1}{b(n+1)})\right)^n +$$

$$\left. (2^n \eta - 1)\alpha + (1-\eta)\alpha\left(1+\exp(-\frac{1}{b(n+1)})\right)^n\right]$$

$$\leq \frac{1}{2b(2^n - 1)} \exp\left(\frac{y}{b}\right) \left[\left(2^n \eta - 1\right)\left((1-\alpha)\exp\left(-\frac{1}{b}\right) + \alpha\right) + \right.$$

$$\left. (1-\eta)(1-\alpha)\left(1+\exp(-\frac{1}{b(n+1)})\right)^n + (1-\eta)\alpha\left(1+\exp(-\frac{1}{b(n+1)})\right)^n\right]$$

$$= \frac{1}{2b(2^n - 1)} \exp\left(\frac{y}{b}\right) \left[\left(2^n \eta - 1\right)\left((1-\alpha)\exp\left(-\frac{1}{b}\right) + \alpha\right) + \right.$$

$$\left. (1-\eta)\left(1+\exp(-\frac{1}{b(n+1)})\right)^n\right]$$

$$\leq \frac{1}{2b(2^n - 1)} \exp\left(\frac{y}{b}\right) \left[2^n \eta\left((1-\alpha)\exp\left(-\frac{1}{b}\right) + \alpha\right) + \right.$$

$$\left. (1-\eta)\left(1+\exp(-\frac{1}{b(n+1)})\right)^n\right].$$

Using $b(n+1) = \frac{1}{\varepsilon}$ to achieve $\varepsilon$-DP, we obtain the following lower bound on $\ell(D_1 \to y)$ with $y \leq 0$:

$$\ell(D_1 \to y) = \log \frac{\max_{d_1 \in \{0,1\}} P_{Y|D_1=d_1}(y)}{P_Y(y)} = \log \frac{P_{Y|D_1=0}(y)}{P_Y(y)}$$

$$\geq \log \frac{2^n \eta + (1+e^{-\varepsilon})^n (1-\eta) - 1}{2^n \eta\left((1-\alpha)\exp(-\varepsilon n - \varepsilon) + \alpha\right) + (1+e^{-\varepsilon})^n (1-\eta)}$$

$$= \log \frac{2^n \eta + (1+e^{-\varepsilon})^n (1-\eta) - 1}{2^n \eta \alpha + \left(\frac{2}{e^\varepsilon}\right)^n \eta(e^{-\varepsilon})(1-\alpha) + (1+e^{-\varepsilon})^n (1-\eta)}.$$

Note that $1 + e^{-\varepsilon} < 2$ and $\frac{2}{e^\varepsilon} < 2$ for all $\varepsilon > 0$. Therefore, when $n$ is large the dominating term in the numerator is $2^n \eta$ and the dominating term in the denominator is $2^n \eta \alpha$. Hence, as $n \to \infty$, the lower bound on $\ell(D_1 \to y)$

approaches $\varepsilon_{\max}(D_1) = \log\frac{1}{\alpha}$. This proves that for each $\delta > 0$, there exists an integer $n$, a database $X$ of size $n$, and a mechanism satisfying $\varepsilon$-DP such that

$$\ell(D_1 \to y) > \varepsilon_{\max}(D_1) - \delta.$$

$\square$

The proof of Theorem 3 relies on a database exhibiting what may be considered as pathologically strong correlations: If the first entry is zero (resp. one) then all other entries are likely to be zero (resp. one) with a constant probability that does not diminish with growing database size $n$. However, it is important to note that the theorem holds true even in more realistic scenarios characterized by weaker correlations. Specifically, the asymptotic lower bound of $\varepsilon_{\max}(D_1)$ for PML remains applicable even if $\eta$ diminishes at a polynomial rate, i.e., if $\eta = \Theta(\frac{1}{n^r})$ for some constant $r \geq 1$.

## 4   Conclusions

While previous research has hinted at the inadequacy of differential privacy in protecting databases with correlated entries, our analysis using pointwise maximal leakage offers a quantitative evaluation of the potential weaknesses in DP's guarantees. Our investigation illuminates the vulnerabilities arising from DP's distribution-agnostic definition, even in its strongest form, i.e., pure DP, which could foster a false sense of privacy if applied indiscriminately. In contrast, PML's ability to adjust to diverse scenarios due to its distribution-dependent nature provides a more nuanced approach to privacy preservation. In this way, our work highlights the necessity for further research aimed at developing a wide array of PML mechanisms tailored to specific contexts.

**Acknowledgments.** This work has been supported by the Swedish Research Council (VR) under the grant 2023-04787 and Digital Futures center within the collaborative project DataLEASH.

**Disclosure of Interests.** The authors have no competing interests to declare that are relevant to the content of this article.

## References

1. Abowd, J.M.: The US Census Bureau adopts differential privacy. In: Proceedings of the 24th ACM SIGKDD International Conference on Knowledge Discovery and Data Mining, pp. 2867–2867 (2018)
2. Alvim, M.S., Chatzikokolakis, K., Palamidessi, C., Smith, G.: Measuring information leakage using generalized gain functions. In: 2012 IEEE 25th Computer Security Foundations Symposium, pp. 265–279 (2012)

3. Bassily, R., Groce, A., Katz, J., Smith, A.: Coupled-worlds privacy: exploiting adversarial uncertainty in statistical data privacy. In: 2013 IEEE 54th Annual Symposium on Foundations of Computer Science, pp. 439–448. IEEE (2013)

4. Bun, M., Steinke, T.: Concentrated differential privacy: simplifications, extensions, and lower bounds. In: Hirt, M., Smith, A. (eds.) TCC 2016. LNCS, vol. 9985, pp. 635–658. Springer, Heidelberg (2016). https://doi.org/10.1007/978-3-662-53641-4_24

5. Cormode, G.: Personal privacy vs population privacy: learning to attack anonymization. In: Proceedings of the 17th ACM SIGKDD International Conference on Knowledge Discovery and Data Mining, pp. 1253–1261 (2011)

6. Ding, B., Kulkarni, J., Yekhanin, S.: Collecting telemetry data privately. Adv. Neural Inf. Process. Syst. **30** (2017)

7. Dong, J., Roth, A., Su, W.J.: Gaussian differential privacy. J. R. Stat. Soc. Ser. B Stat Methodol. **84**(1), 3–37 (2022)

8. Dwork, C.: Differential privacy. In: Bugliesi, M., Preneel, B., Sassone, V., Wegener, I. (eds.) ICALP 2006. LNCS, vol. 4052, pp. 1–12. Springer, Heidelberg (2006). https://doi.org/10.1007/11787006_1

9. Dwork, C., Kenthapadi, K., McSherry, F., Mironov, I., Naor, M.: Our data, ourselves: privacy via distributed noise generation. In: Vaudenay, S. (ed.) EUROCRYPT 2006. LNCS, vol. 4004, pp. 486–503. Springer, Heidelberg (2006). https://doi.org/10.1007/11761679_29

10. Dwork, C., McSherry, F., Nissim, K., Smith, A.: Calibrating noise to sensitivity in private data analysis. In: Halevi, S., Rabin, T. (eds.) TCC 2006. LNCS, vol. 3876, pp. 265–284. Springer, Heidelberg (2006). https://doi.org/10.1007/11681878_14

11. Dwork, C., Naor, M.: On the difficulties of disclosure prevention in statistical databases or the case for differential privacy. J. Privacy Confident. **2**(1) (2010)

12. Dwork, C., Roth, A., et al.: The algorithmic foundations of differential privacy. Found. Trends® Theor. Comput. Sci. **9**(3–4), 211–407 (2014)

13. Dwork, C., Rothblum, G.N.: Concentrated differential privacy. arXiv preprint arXiv:1603.01887 (2016)

14. Dwork, C., Smith, A., Steinke, T., Ullman, J.: Exposed! a survey of attacks on private data. Annu. Rev. Stat. Appl **4**(1), 61–84 (2017)

15. Erlingsson, Ú., Pihur, V., Korolova, A.: Rappor: randomized aggregatable privacy-preserving ordinal response. In: Proceedings of the 2014 ACM SIGSAC Conference on Computer and Communications Security, pp. 1054–1067 (2014)

16. He, X., Machanavajjhala, A., Ding, B.: Blowfish privacy: tuning privacy-utility trade-offs using policies. In: Proceedings of the 2014 ACM SIGMOD International Conference on Management of Data, pp. 1447–1458 (2014)

17. Issa, I., Wagner, A.B., Kamath, S.: An operational approach to information leakage. IEEE Trans. Inf. Theory **66**(3), 1625–1657 (2019)

18. Kifer, D., Machanavajjhala, A.: No free lunch in data privacy. In: Proceedings of the 2011 ACM SIGMOD International Conference on Management of Data, pp. 193–204 (2011)

19. Kifer, D., Machanavajjhala, A.: A rigorous and customizable framework for privacy. In: Proceedings of the 31st ACM SIGMOD-SIGACT-SIGAI Symposium on Principles of Database Systems, pp. 77–88 (2012)

20. Kifer, D., Machanavajjhala, A.: Pufferfish: a framework for mathematical privacy definitions. ACM Trans. Datab. Syst. **39**(1), 1–36 (2014)

21. Li, N., Qardaji, W., Su, D., Wu, Y., Yang, W.: Membership privacy: a unifying framework for privacy definitions. In: Proceedings of the 2013 ACM SIGSAC Con-

ference on Computer and Communications Security - CCS 2013, pp. 889–900. ACM Press, Berlin (2013)

22. Liu, C., Chakraborty, S., Mittal, P.: Dependence makes you vulnberable: differential privacy under dependent tuples. In: NDSS, vol. 16, pp. 21–24 (2016)

23. Mironov, I.: Rényi differential privacy. In: 2017 IEEE 30th Computer Security Foundations Symposium (CSF), pp. 263–275. IEEE (2017)

24. Rényi, A.: On measures of entropy and information. In: Proceedings of the Fourth Berkeley Symposium on Mathematical Statistics and Probability, Berkeley, vol. 1, pp. 547–561 (1961)

25. Saeidian, S., Cervia, G., Oechtering, T.J., Skoglund, M.: Pointwise maximal leakage. IEEE Trans. Inf. Theory **69**(12), 8054–8080 (2023)

26. Saeidian, S., Cervia, G., Oechtering, T.J., Skoglund, M.: Pointwise maximal leakage on general alphabets. In: 2023 IEEE International Symposium on Information Theory (ISIT), pp. 388–393. IEEE (2023)

27. Saeidian, S., Cervia, G., Oechtering, T.J., Skoglund, M.: Rethinking disclosure prevention with pointwise maximal leakage. Submitted to: Journal of Privacy and Confidentiality. (2023). https://people.kth.se/~oech/JPC23.pdf

28. Thakurta, A.G., et al.: Learning new words (14 March 2017), uS Patent 9,594,741

29. Warner, S.L.: Randomized response: a survey technique for eliminating evasive answer bias. J. Am. Stat. Assoc. **60**(309), 63–69 (1965)

30. Yang, B., Sato, I., Nakagawa, H.: Bayesian differential privacy on correlated data. In: Proceedings of the 2015 ACM SIGMOD international conference on Management of Data. pp. 747–762 (2015)

31. Zhu, T., Xiong, P., Li, G., Zhou, W.: Correlated differential privacy: hiding information in non-IID data set. IEEE Trans. Inf. Forens. Secur. **10**(2), 229–242 (2014)

# Addressing Privacy Concerns in Joint Communication and Sensing for 6G Networks: Challenges and Prospects

Prajnamaya Dass[1]([✉]), Sonika Ujjwal[2], Jiri Novotny[2], Yevhen Zolotavkin[1], Zakaria Laaroussi[2], and Stefan Köpsell[1]

[1] Barkhausen Institute, Dresden, Germany
{prajnamaya.dass,yevhen.zolotavkin,
stefan.koepsell}@barkhauseninstitut.org
[2] Oy LM Ericsson Ab, Jorvas, Finland
{sonika.a.ujjwal,jiri.novotny,zakaria.laaroussi}@ericsson.com

**Abstract.** The vision for 6G extends beyond mere communication, incorporating sensing capabilities to facilitate a diverse array of novel applications and services. However, the advent of joint communication and sensing (JCAS) technology introduces concerns regarding the handling of sensitive personally identifiable information (PII) pertaining to individuals and objects, along with external third-party data and disclosure. Consequently, JCAS-based applications are susceptible to privacy breaches, including location tracking, identity disclosure, profiling, and misuse of sensor data, raising significant implications under the European Union's general data protection regulation (GDPR) as well as other applicable standards. This paper critically examines emergent JCAS architectures and underscores the necessity for network functions to enable privacy-specific features in the 6G systems. We propose an enhanced JCAS architecture with new network functions and interfaces, facilitating the management of sensing policies, consent information, and transparency guidelines, alongside the integration of sensing-specific functions and storage for sensing processing sessions. Furthermore, we conduct a comprehensive threat analysis for all interfaces, employing security threat model STRIDE and privacy threat model LINDDUN. We also summarise the identified threats using standard common weakness enumeration (CWE). Finally, we suggest the security and privacy controls as the mitigating strategies to counter the identified threats stemming from the JCAS architecture.

**Keywords:** JCAS · Joint communication and sensing · ISAC · ICAS · Integrated communication and sensing · 6G · Threats · Privacy · Security

## 1 Introduction

The evolution of mobile communication, spanning from the inception of 1G to the latest iteration 5G, has reshaped the fabric of human connectivity and inter-

M. Jensen et al. (Eds.): APF 2024, LNCS 14831, pp. 87–111, 2024.
https://doi.org/10.1007/978-3-031-68024-3_5

action. While technologies like beamforming and network slicing have bolstered efficiency, 5G remains primarily communication-focused [35]. Looking ahead, 6G aims to surpass mere communication, integrating sensing for a myriad of innovative applications and services [38].

The integration of communication and sensing capabilities, also referred as joint communication and sensing (JCAS), reflects a growing enthusiasm to unlock its capabilities in solving real-life challenges efficiently. In the literature, the term JCAS is alternatively denoted as integrated communication and sensing (ICAS) or integrated sensing and communication (ISAC). Throughout this paper, we use the term JCAS to represent this concept. There are many use cases proposed by 3GPP and other organisations that focus on the potential applications of JCAS across various domains, e.g., autonomous driving, smart city, precision agriculture, industrial IoT, healthcare and telemedicine [2,38]. Despite its numerous technological benefits and emerging use cases, JCAS introduces various security and privacy challenges [36,40]. Given the incorporation of sensing data, which may contain sensitive personally identifiable information (PII) regarding individuals and objects, JCAS-based applications are increasingly vulnerable to privacy attacks, including location tracking, identity disclosure, profiling, and misuse of sensor data.

In the context of JCAS, the predominant emphasis in current security and privacy solutions lies within physical layer. On the other hand, the privacy mechanisms tailored for independent sensing mechanisms may not be suitable for JCAS scenarios and their particular use cases. The integration of sensing into the current 3GPP architecture necessitates supplementary core network functionalities beyond the sensing management function (SeMF). Additionally, given these added core functions and changes in radio signalling, a distinct threat analysis is imperative. Consequently, the mitigation strategies should include strong security and privacy measures to effectively counter these specific threats posed by JCAS. Therefore, recognising the sensitivity of JCAS technology and the gaps in existing literature, we present the following contributions in this study.

- We conduct a critical assessment of the emergent JCAS architecture, identifying potential security and privacy challenges (Sect. 3);
- To tackle these challenges, we propose enhancements to the emergent JCAS architecture, introducing new network functions (NFs) and interfaces (Sect. 4);
- We perform a comprehensive threat analysis of the proposed JCAS architecture, considering the introduced interfaces and components. Our analysis utilises both STRIDE and LINDDUN threat models to cover security and privacy risks comprehensively (Sect. 5);
- Finally, we suggest security and privacy controls to counter the identified threats (Sect. 6).

## 2    Related Works

Due to the widespread interest in JCAS, researchers have extensively explored various facets of this paradigm. Architectural concepts, primarily addressing anticipated alterations in the core network functionalities to incorporate sensing alongside communications, have been discussed in works [26,39,45].

Many existing works focus on JCAS security at the physical layer. In [14], the study addresses the challenge of reducing information leakage between communication and sensing functionalities within systems that concurrently perform both operations. In [34], the authors conduct a comprehensive security assessment of spoofing attacks in an mmWave radar-based sensing system for autonomous vehicles, incorporating the development and execution of tangible physical layer attack and defence tactics within a cutting-edge mmWave test environment. A spatio-temporal spoofing detection mechanism leveraging MIMO beamforming, was proposed in [20] to mitigate spoofing against automotive radars.

In [27], the authors emphasised privacy issues due to sensing activities, such as activity monitoring of sensing targets, eavesdropping attacks from the sensing signals, and false data injection attacks. They suggest the creation of a cross-domain technique for sensing and localisation to accurately recognise human activities, becoming less dependent on location. The work [21], explores privacy concerns surrounding personal sensing, where individuals utilise devices to monitor their activity, location, and environment, and proposing strategies to enhance privacy sensitivity in personal sensing technologies. Participatory sensing allows users to collect and share data through their mobile devices. The work [8], investigates the privacy concerns due to the use of multi-modal sensors in mobile phones, evaluates existing privacy solutions, and discusses potential countermeasures to safeguard user privacy. In similar works [10,12], the authors addresses privacy protection by defining requirements, proposing an efficient infrastructure for mobile users, and discussing open problems and research directions.

## 3   Emergent JCAS Architecture

Aligning with the architectural concepts outlined in [26,39,45], we use the emergent JCAS architecture depicted in Fig. 1 as a baseline for further discussion. To deal with the sensing activities, an additional network function – sensing management function (SeMF), sometimes referred to as Sensing function (SF) is added to the core network. The SeMF has two main components, namely, sensing control function (SCF) and sensing processing function (SPF). SCF leverages the control plane to receive sensing requests and orchestrate necessary actions with other network functions and entities, such as sensing enabled base stations (gNBs) and UEs. Sensing requests originating in an application (APP) are authenticated and authorised by the network exposure function (NEF). Alternatively sensing requests could be invoked directly from an application function (AF). After sensing signalling between relevant gNBs and UEs the collected sensing measurements are reported to SPF, using for example data plane [26], for processing to gain a semantic understanding of the physical environment depending on the sensing task. SPF may perform tasks such as data aggregation, signal processing, object classification, and anomaly detection to enhance situational awareness and provide results to the application through the NEF.

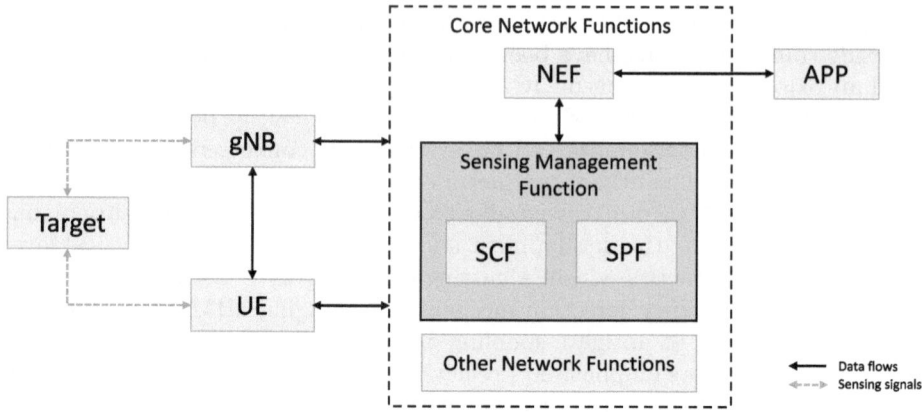

**Fig. 1.** Simplified depiction of the emergent JCAS architecture based on prior art.

### 3.1 Privacy and Security Challenges in the Emergent JCAS Architecture

The challenge of a privacy-preserving JCAS architecture remains unresolved, as the nature of sensing data differs from communication data, posing difficulties in integrating sensing into existing 5G/5G-Advanced frameworks. Unlike communication data, sensing data, akin to other sensor data, primarily captures observations about the physical environment and its objects, which may not always be directly linked to a subscriber. For instance, in JCAS-assisted automotive manoeuvring and navigation system, a vehicle user may be tracked by other vehicles and the service provider. The service provider may control the sensing unit of the vehicle and use other vehicle features without user's consent. On the other hand, a vehicle with sensing capabilities can also sense the targets in sensitive or restricted zones. Considering that sensing data may potentially include personal data, whether directly or indirectly, the integration of JCAS in 6G should adhere to the principles outlined in the GDPR on how to collect, use, transfer, store, and dispose the sensing data.

Introducing the additional sensing functions into the core network for JCAS poses several threats. Firstly, it expands the attack surface of the network, providing more avenues for malicious actors to exploit vulnerabilities. Secondly, the increased volume and diversity of sensing data collected through these sensing functions heighten the risk of data breaches and privacy violations.

Additionally, attackers could exploit vulnerabilities in one sensing function to breach trust boundaries and gain unauthorised access to other network components. Inadequate enforcement of trust boundaries may lead to data leakage between sensing functions or network domains, compromising the confidentiality and integrity of sensitive information. Robust access controls, authentication mechanisms, and continuous monitoring are essential to mitigate threats associated with trust boundary violations and ensure the security of the JCAS system. Addressing these threats requires robust security measures, stringent privacy protections, and effective management practices to ensure the resilience and integrity of the JCAS system.

## 4  Proposed Architectural Enhancement

In this section, we describe the architecture that is evolving from the prior art architecture from Sect. 3 and introducing new concepts in attempts to address some of the privacy challenges covered in Sect. 3.1.

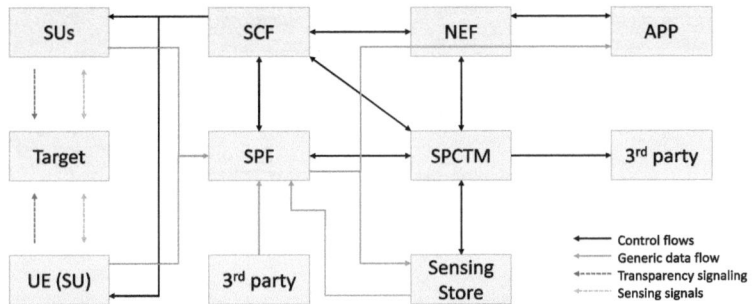

**Fig. 2.** Bird's-eye view of the assumed data flow diagram for this work.

Figure 2 shows a bird's-eye view of the proposed architecture. Similarly to prior art, the NEF brings network capabilities to applications and we retain the SCF and the SPF. However, we propose a complementing component called the sensing policy, consent, and transparency management (SPCTM), which governs sensing privacy, as well as a sensing store to hold persistent data, policies, and logs tailored to the requirements of the sensing function. We propose that the sensing store is an NF-specific solution at this stage, which is not uncommon in current 5G deployments [1], in order to grasp requirements and possible threats against such an approach. Future studies may investigate the possibilities of reusing existing data store solutions in 5G architecture such as UDM, UDR, UDSF etc. We use the term sensing unit (SU) to refer to the sensing radio component that can be independent (e.g., at the gNB) or part of the user equipment (UE) [2]. Note that in the latter case, additional UE-based controls govern access to the sensing data to put the user under control. SUs transmit sensing signals to the targets and then capture the reflected signals from them, additionally SUs may be instructed to notify sensing targets about the current sensing session over an air interface such as broadcast. Captured sensing signals are interpreted into sensing measurements which are then sent to the SPF for further processing. The processed measurements, named sensing results, are then disclosed via the NEF back to the application. The interfaces needed towards communication system, depending on the level of integration [41], are not part of the scope of this document. The working assumptions and detailed descriptions of each component are provided in the following subsections.

## 4.1   The Network Exposure and Application

The NEF is an essential part of 5G architecture, enabling secure third-party access to 3GPP network services and capabilities, while enforcing policies on data sharing through APIs [3]. Figure 3 shows how the NEF enables sensing services, including authentication and policy-based authorisation of sensing applications, with help from the sensing authorisation and policies check component.

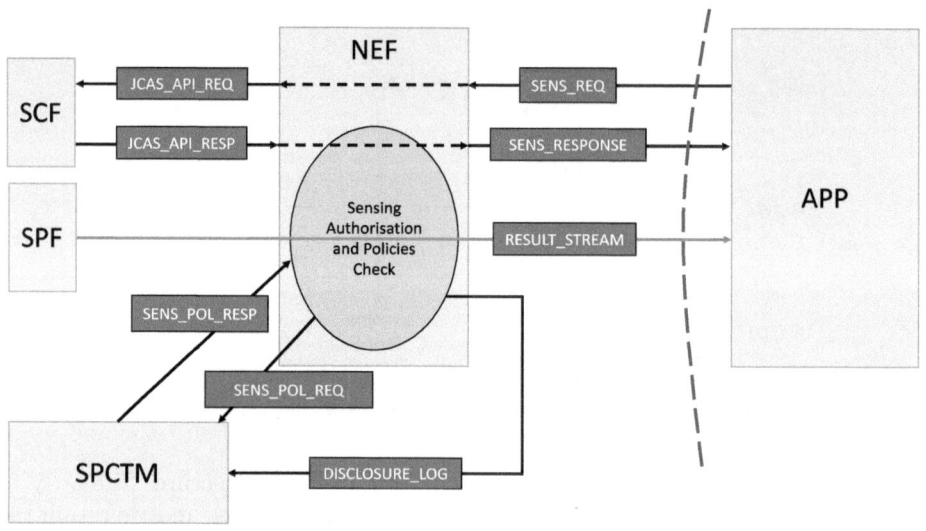

**Fig. 3.** Proposed control and data flows from and to the NEF.

Sensing is initiated with the sensing request (SENS_REQ), which at the very least contains descriptions of the sensing target and sensing results, which differ depending on the given use case. For example, for an early collision warning application on a highway [2], this could be the geo-location as the target and an event-like notification as the response. However, depending on the scenario, the request is expected to be elaborate, containing other fields such as quality of service (QoS), quality of data (QoD), periodicity, and more. In return, the application receives a sensing response (SENS_RESPONSE) indicating the status - success or failure. The response may carry a sensing result or provide information on how to obtain the results, for example, a web socket to listen in for stream-like sensing results or events (RESULT_STREAM).

The NEF requests necessary policies (SENS_POL_REQ) from the SPCTM, receiving sensing specific authentication and authorisation details (SENS_POL_RESP) for the sensing application, such as geo-location permissions and result granularity. The SPCTM assigns a reference (e.g., policy ID) for policy tracking, shared across the NEF, SCF, and SPF. These data flows occur initially and may repeat periodically to ensure proper disclosure of sensing responses and results to the application. The NEF records data disclosures for transparency, complying with legal standards like GDPR via the DISCLOSURE_LOG flow. It logs for example recipient identities, data descriptions, disclosure purposes, obligations, timestamps, and applied policies.

After the sensing request has been authorised and the initial set of policies has been established, the NEF proceeds to relay the request to the SCF using a lower-level style API (JCAS_API_REQ). In response (JCAS_API_RESP), the NEF receives a detailed message from the SCF indicating whether the request was successfully processed or if it encountered a failure.

## 4.2 Sensing Policy, Consent, and Transparency Management (SPCTM)

The SPCTM function is designed to administer privacy controls and extend support to other NFs participating in the sensing ecosystem, enabling them to adhere to privacy preservation principles. The SPCTM framework proposed herein does not encompass the privacy considerations for all conceivable use cases; instead, it proposes a foundational model.

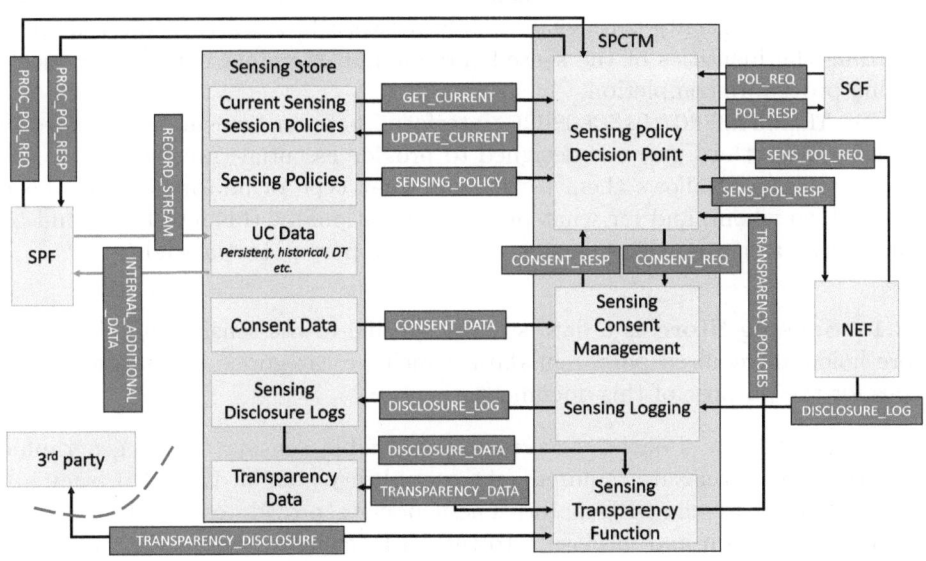

**Fig. 4.** SPCTM, governing sensing privacy, and Sensing Store interactions with each other and the rest of the proposed system.

The components and interfaces of the SPCTM are illustrated in Fig. 4. The sensing policy decision (SPD) point functions as a central hub for gathering and consolidating sensing policies, consent information, and transparency guidelines, and subsequently disseminating this aggregated information to other NFs. It also maintains a record of the current policies applicable to active sensing sessions. It is presupposed that these policies and associated consent or transparency data, although predetermined, may be subject to change over the course of a sensing session. The SPD point is tasked with the timely notification and updating of relevant components and NFs to reflect these changes, acting as the primary interface between the SPCTM and the remainder of the system. Furthermore, the SPD point is charged with negotiating current privacy policies with the various NFs involved in the sensing process.

The sensing consent management component bears the responsibility for handling consent data and supplying it to the SPD point. Recognising the necessity of such a component is vital for supporting a broader array of sensing applications. However, the precise technical methods for obtaining and managing the consent of all stakeholders are topics for further studies.

The Sensing Logging function interfaces with the NEF, as previously outlined in Sect. 4.1, to facilitate disclosure logging that adheres to transparency requirements. These requirements are provided by the sensing transparency function, which is also responsible for disseminating information on how sensing sessions should be communicated to the affected sensing targets. Options under consideration include directing the SCF, which then appoints SUs, to emit a transparency notification, potentially through a mobile network broadcast, or alternatively, recording the identities of the sensed targets and providing notifications post-sensing procedure completion.

The TRANSPARENCY_DISCLOSURE interface, between sensing transparency function and a third party, is designed to provide essential transparency to the sensing targets. It allows them to observe what type of information has been disclosed, to whom, and for what purpose. Additionally, this interface could be managed by a trusted third party or used to support potential audits.

**4.2.1    Sensing Store** is a data storage specific to the sensing function. This store holds all required persistent data, which we categorise into the following types for the purpose of this document.

1. Sensing Policies: Policies governing permissible sensing types, geographic restrictions, disclosure requirements, granularity standards, and privacy are consolidated by the SPD point. These policies provide essential guidelines for authorisation and disclosure to the NEF, control for the SCF, and data handling for the SPF.
2. Consent Data: This includes consents from sensing targets, where consent is the legal basis for data collection. It covers user permissions for sensing activities on their devices and the extent of their participation. Managing these consents is complex due to the indirect, potentially sensitive, and large-scale nature of sensing data.
3. Current Sensing Session Policies: These reflect the latest aggregated policies for authorisation, disclosure, control, and processing in active sessions. The

storage facilitates access to relevant policies, consent data, and session information, along with any policies composed by the SPD point.

4. Use Case (UC) Data: Persistent data storage for a specific use case, such as environment maps or historical records, should exclude PIIs and contain only sanitised data. Sensitive data management is addressed separately within the SPF (Sect. 4.5.1).

5. Sensing Disclosure Logs: Records of data disclosure detail recipient identities, data descriptions, disclosure purposes, obligations, timestamps, and the policies enforced during the process.

6. Transparency Data: Policies outline how the collection and processing of sensing data are communicated to impacted individuals and how disclosure is logged and shared with relevant parties, aligning with the mentioned TRANSPARENCY_DISCLOSURE interface.

### 4.3   The Sensing Control and Orchestration

Similar to [26,39], the SCF operates within the control plane of the JCAS framework, as illustrated in Fig. 5. The internal API of the SCF facilitates the flow of sensing requests (JCAS_API_REQ). The SCF aggregates multiple requests that can be fulfilled within a single procedure or session, or segregates a single request that cannot be fulfilled within one session, into the necessary set of sensing tasks for measurements and processing.

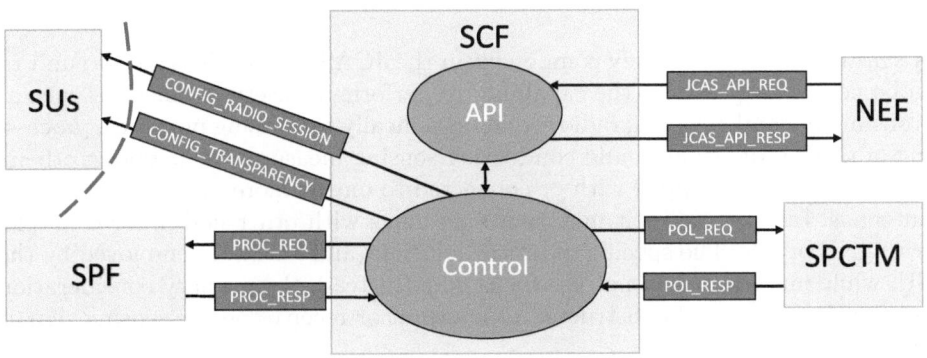

**Fig. 5.** Control flows associated with SCF for managing and orchestrating other NFs to provide requested sensing results.

Upon successful negotiation with the SPCTM over the control policies (via POL_REQ and POL_RESP), the SCF requests processing resources for the given task from the SPF using a processing request (PROC_REQ). The PROC_REQ includes an estimate of the type, size, and frequency of incoming data, priority or criticality levels, and other necessary parts for processing related to the results required (type, periodicity, reporting style - single, stream, or event) and references to current policies. The PROC_RESP is a processing response that includes a success or failure indicator, ingress points definitions (e.g., IP address), and optionally, egress points definition (IP address, socket). SCF then sends control parameters to the relevant SUs using the CONFIG_RADIO_SESSION command. The SUs execute the sensing operation and send the raw or optionally pre-processed sensing data back to the SPF. The SCF also manages the trade-off between communication and sensing services by efficiently allocating resources. It can potentially fail a request if sensing measurements cannot be obtained as desired, or if the SPF fails to secure resources for needed processing.

The introduction of SPCTM supports more heterogeneous aggregation of sensing tasks by applying correct policies and resolving potentially conflicting ones. The POL_RESP data contains the current set of control-related policies such as granularity recommendations (time and space), transparency signal information and more. The communication between the SPF and SPCTM is envisioned as a negotiation sequence, allowing the SCF and the SPD point to agree on a solution. The SCF needs to bundle requests together, and the SPD point needs to support this by bundling and resolving relevant policies. Furthermore, the SPD point may need to reflect changes in control policies to processing policies. For example, if a higher granularity is used for sensing measurements, the processing pipeline should compensate as soon as technologically possible. The received indications regarding transparency signalling in POL_RESP towards affected sensing targets are instructed in CONFIG_TRANSPARENCY to SUs.

## 4.4 Sensing Units

A sensing unit (SU) is a key component in the JCAS system. It is a radio unit or radio node that possesses the capability to perform a variety of functions such as transmitting and receiving radio signals specifically for sensing purposes, processing of these radio signals, and conducting sensing measurements, among others. The SU may be equipped with or connected to one or more internal or external antennas. In some cases, it may share antennas with other nodes, for example, with gNB or UE. The specific technical solutions and hardware employed by the SU, while important, are not the focus here. Instead, the primary consideration is the SU's ability to be instructed to transmit or receive specific sensing signals at specific times.

As illustrated in Fig. 6, the scope is extended to UE-based SUs. Note that these SUs are not directly controlled by the network. Instead, the UE will implement similar privacy controls as the ones in the network (especially SPCTM) (see Sect. 6.2.4 for more details). Therefore only indirect control is possible, which is governed by user decisions and policies. The controls are omitted in Fig. 6 for readability. They are presented in Fig. 9.

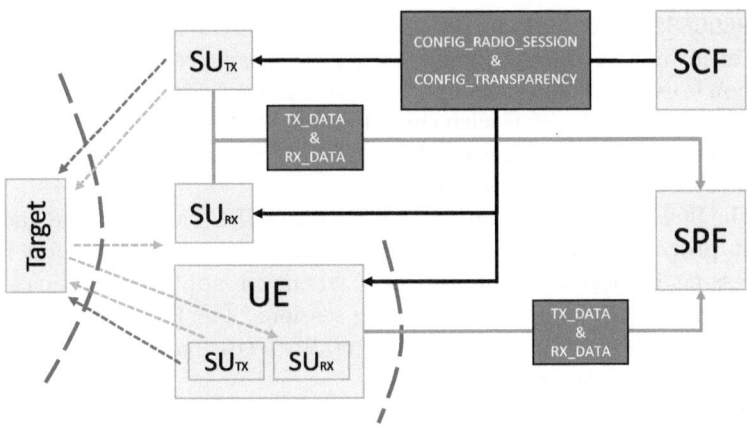

**Fig. 6.** Control flows (CONFIG_RADIO_SESSION and CONFIG_TRANSPARENCY) from the SCF to SUs and data flows (TX_DATA and RX_DATA) from SUs to the SPF.

### 4.5    Sensing Processing Function

The SPF handles raw and pre-processed sensing data from SUs, converting it into usable results for applications delivered via the NEF. As shown in Fig. 7, SPF includes a control component that responds to processing requests (PROC_REQ) from the SCF, which provides necessary details for processing the data. This component orchestrates processing sessions, managing the required resources for computation and storage to meet the request.

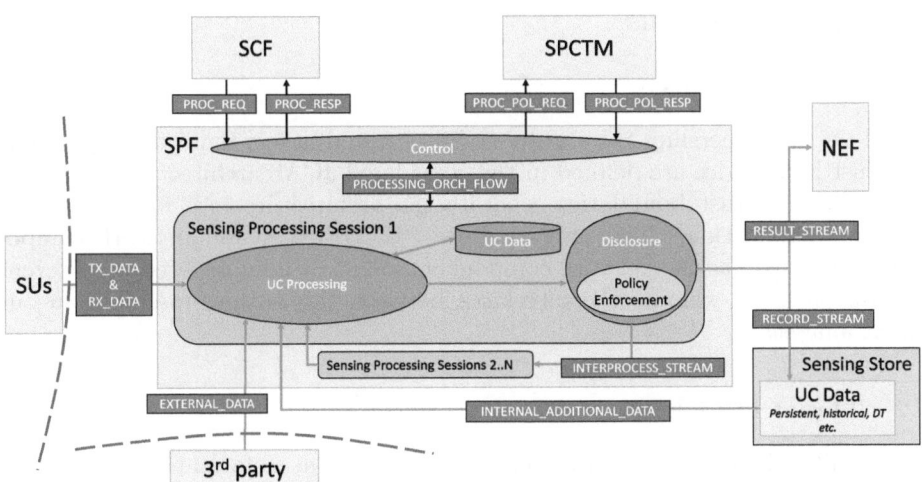

**Fig. 7.** Control flows between SPF, SCF, and SPCTM for sensing data processing.

The `PROCESSING_ORCH_FLOW` carries all the orchestration information for SPF sessions, ensuring the processing aligns with the sensing request or task. Once the SPF session is established, it sets up the ingress point for incoming data. The SPF control then updates the SCF with the fulfilment status in the `PROC_RES` message. If successful, it also relays ingress (e.g., IP address and port) and potentially egress point details for accessing results.

Like the SCF, the SPF communicates with the SPCTM to receive and enforce the latest processing policies. This interaction occurs initially upon receiving the processing request and subsequently whenever applicable policies change, requiring adjustments in active processing sessions. The `PROCESSING_ORCH_FLOW` may include directives to initiate, end, or modify a session in response to changes in sensing or processing policies.

**4.5.1  Sensing Processing Sessions** are transient instances providing necessary processing for fulfilling sensing requests and which can share intermediate data via the `INTERPROCESS_STREAM` interface among themselves, as shown in Fig. 7. Each session, with its temporary data store and processes, is self-contained and securely disposes of sensitive data upon completion at very least. Depending on privacy needs, sessions may operate in secure environments such as confidential VMs, enclaves, or containers. Upon receiving `TX_DATA` and `RX_DATA`, SPF sessions aggregate data for a unified view, then process it for specific requests and use cases, such as adding semantic information about the environment. The UC data storage holds temporary data and interim results. The disclosure component ensures privacy controls are in place before releasing results to the app via the NEF or transferring to other sessions.

## 5  Threat Analysis

### 5.1  Trust Domains

Based on the ownership, roles, responsibilities, and access requirements, different trust boundaries are defined in the considered JCAS architecture, as shown in Fig. 8. The trust boundaries separate the architecture into following trust domains: application, third party, target, SU, UE(SU), and network components. Moreover, as the network components might be administered by different entities, we also examine the interfaces between the sensing functions in our threat analysis.

### 5.2  Threat Models

Both STRIDE (Spoofing, Tampering, Repudiation, Information disclosure, Denial of service (DoS), Elevation of privilege) [17] and LINDDUN (Linkability, Identifiability, Non-repudiation, Detectability, Data disclosure, Unawareness and unintervenability, Non-compliance) [11] models have been utilised for security

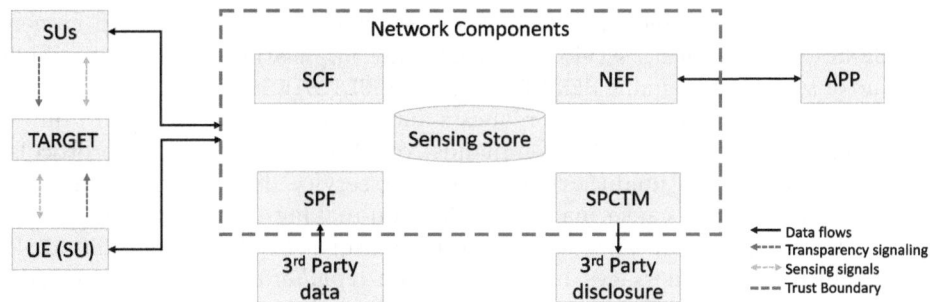

**Fig. 8.** JCAS components and interfaces considered for threat analysis.

and privacy threat analysis, leveraging their widespread adoption and systematic threat identification in conceptual architecture modelling. While STRIDE focuses on security threats compromising confidentiality, integrity, or availability, LINDDUN examines potential privacy implications associated with personally identifiable information during data processing activities. The selected threats for analysis include spoofing, tampering, repudiation, information disclosure, denial of service, elevation of privilege, linkability, identifiability, detectability, unawareness, and non-compliance.

### 5.3   Threats

Appendix A in [9] provides a summary of the threats in the proposed JCAS architecture, considering the interfaces discussed in Sect. 4. We specify the standard common weakness enumeration (CWE) [28] in the threat table (please refer to Appendix A in [9]) to enhance clarity and precision by providing standardised identification of weaknesses associated with each threat. The subsequent discussions give detailed analysis of the threats associated with each interface. Initially, our analysis focuses on external interfaces to assess potential threats. Then, we evaluate threats from intra-network interfaces, considering distinct trust boundaries for each network function.

**5.3.1   Threats in Application ⟷ NEF Interface** Due to the presence of sensing responses, the interface between NEF and applications poses a potential target for diverse threats. *Linkability* threats may emerge if attackers can correlate multiple SENS_REQ and SENS_RESPONSE exchanges, potentially exposing sensitive behaviours or patterns. Insufficient security measures for protecting authentication tokens and session identifiers exchanged between the NEF and the application pose risks of *identifiability* and *spoofing*. Moreover, interception of PIIs of the NEF could allow attackers to impersonate the NEF and send malevolent responses to applications. Improper data protection mechanisms may expose sensitive sensing data in the RESULT_STREAM to eavesdropping.

Attackers may conduct *DoS* attacks by overwhelming the NEF with numerous SENS_REQ, disrupting services for legitimate applications. Similarly, a high volume of seemingly genuine SENS_RESP and RESULT_STREAM could impact applications. By manipulating sensing requests and responses, attackers can tamper with control flows, inject false commands, or alter communication protocols, potentially leading to unauthorised actions or service degradation. Tampered RESULT_STREAM could cause inaccurate decision-making or operational disruption. Additionally, tampering with digital signatures in SENS_REQ, SENS_RESP, and RESULT_STREAM messages undermines *non-repudiation*, allowing attackers to repudiate legitimate transactions if private keys are compromised.

If access control policies are inadequately designed, an application with an *elevation of privilege* can request sensing data from unauthorised environments or restricted areas like military zones. Failure to communicate policy changes, such as data categorisation or PII classification, to the application and NEF, may inadvertently disclose sensitive information in sensing requests or responses. *Non-compliance* and policy violations may grant attackers access to sensitive data from application sensing sessions, potentially involving legitimate NEF and application participation in illegitimate sensing activities.

**5.3.2    Threats in SU/UE(SU) ⟷ Network Interface** The wireless nature of this interface renders it susceptible to numerous threats. Attackers could *link* the sensing environment, units involved, and requesting applications from exchanged radio configuration and sensing information. Inadequate confidentiality measures may *disclose* sensitive data like sensed targets, areas, involved units, and network ingestion points. Furthermore, session identifiers used for radio session configuration may provide *identifiable* information about entities participating in specific sessions.

By *spoofing* the PIIs of the SCF, an attacker may send unauthorised radio configuration requests CONFIG_RADIO_SESSION to SUs and collect sensing data. Alternatively, if PIIs of the SU are disclosed, the attacker can send malicious data and acknowledgements to the network, potentially ensuring non-repudiation to appear legitimate. *Elevated privileges* could allow the attacker or SCF to control SU sensing sessions and direct them for unlawful activities. *Tampering* with sensing data and injecting malicious programs can disrupt final results and network components. Attackers may conduct *DoS* attacks by flooding the sensing unit with seemingly legitimate CONFIG_RADIO_SESSION messages or jamming to block them, preventing radio sensing initiation. If the SU is integrated into the gNB, it can influence the communication system, including signalling processes.

**When SU Resides in UE:** If the SU is part of the UE, other forms of information besides captured signals from targets are exchanged in the interface, assuming that the UE can process sensing information. Metadata can *disclose* details about sensing sessions, allowing adversaries to link sessions from the application with metadata transferred from the UE. By *elevating privileges*, the SCF may enable sensitive features like GPS and IP localisation of the UE, potentially leading to unauthorised tracking. With *elevated privileges*, the UE may unknowingly engage in sensing activities, necessitating the imposition of sensing policies and regulations. Additionally, standards for handling PIIs should be followed and properly communicated to the UE. Due to the UE's involvement in various processing tasks, the SCF can obtain metadata about the UE's storage and processing capabilities.

### 5.3.3 Threats in Target ⟷ SU Interface

The sensing units emit signals and capture reflections from targets, allowing adversaries to establish connections and potentially *disclose* sensitive information about target movements or activities. Additionally, adversaries might uniquely *identify* or *link* individual targets from observations, jeopardising target anonymity and operational security. Attackers could *impersonate* legitimate targets or deceive sensing units with false signals, leading to misidentification or inaccurate target tracking. Threats during sensing operations may include *DoS* attacks aimed at disrupting radar functionality, lack of awareness or *non-compliance* with access control measures, and violations of privacy regulations, compromising target security and privacy. Similarly, failure to comply with privacy regulations or ethical guidelines may lead to unauthorised collection, storage, or sharing of personal or sensitive information, violating privacy rights.

The integration of sensing units into UEs amplifies privacy concerns, as UEs with sensing capabilities may inadvertently gather sensitive data about individuals or their surroundings. For instance, SU-enabled UEs could potentially monitor users' movements or activities, heightening privacy risks if this data is misused or accessed without authorisation. Conversely, UEs with integrated sensing functionality could be exploited for unauthorised surveillance by malicious actors, presenting privacy threats to the UEs. Attackers could also correlate UEs and their movements with sensed areas or targets during specific sessions, as well as the associated application and network information.

### 5.3.4 Threats in Network ⟷ Third-Party Interface

The incoming malicious or erroneous third-party data may be misleading for the final sensing outcomes of the network and with intention of tampering, the third party could disrupt the network services. Adversaries might launch *DoS* attacks targeting the network interfaces to the third parties, flooding with excessive data or causing network congestion. On the other hand, the network can also do such type of attack by sending malicious information to the third parties.

Security weaknesses in interfaces or integration points between the network and third-party data sources may be exploited by attackers to gain unauthorised access to sensitive data, manipulate sensor readings, and inject malicious content. From the interface, the adversary could get some *identifiable* information and *link* the third parties involved with the network, type of sensing data shared by a third party, and some PIIs for sensing processing session. Failure to comply with data protection regulations, privacy laws, or industry standards governing the handling and sharing of third-party data may result in inadequate security measures, leading to data breaches or unauthorised access.

**5.3.5  Threats in NEF ⟷ SCF Interface** In this interface, a malicious entity can *pretend* to be NEF and send sensing request to SCF if insecure and improper authentication policies are used. Additionally, unauthorised modification can be done to sensing request and response if sufficient mechanisms are not implemented to protect the integrity of communication. Further, an adversary can *observe* the communication and can find out sensitive information about sensing request, such as which entity (APP) is requesting sensing, what is the granularity, and target area of sensing etc.

The adversary can also observe the attributes of *REQ* and *RESP* messages and potentially *link* or, in the worst case, *identify* the entities requesting sensing, sensing targets, and network components involved in sensing. *DoS* threats are also applicable to NEF and SCF communication, as an adversary can *tamper* with the JCAS_API_REQ so that SCF is unable to process the sensing request. This can be achieved either through unsupported fields and parameters or by creating a very large sensing request. Volumetric DoS attacks are also possible if the adversary attempts to flood SCF and NEF with a massive number of JCAS_API_REQ and JCAS_API_RESP messages, respectively. It should be noted that improper access control on either side can result in an entity being able to access resources that it is not authorised to use. It is important for NEF to work in *compliance* with standards and policies and not initiate any unauthorised sensing requests. For example, malicious insider or operator may conduct non-compliant sensing for financial or political motives.

**5.3.6  Threats in SCF ⟷ SPF Interface** An adversary can *spoof* SCF and request SPF to process a sensing request, initiating unauthorised actions such as scan area requests or object tracking. Without secure communication between SCF-SPF, message *tampering* is possible, allowing the adversary to alter critical parameters like ingress points (port address where SU should send the sensing data) and sensing processing information (e.g., size, duration, frequency) etc. *DoS* threats on the SCF-SPF communication channel can occur through crafting large PROC_REQ or flooding with numerous requests of this type, filling up SPF's capability to serve these requests and exhausting its resources. *Disclosed* information from insecure messages, like network ingestion points or sensing processing characteristics, can aid subsequent attacks.

An entity at SCF can *elevate its privileges* to access SPF resources, including SPF sessions, SPF ephemeral data store, and processing pipeline, allowing it to arbitrarily start and end a sensing processing session if proper access control mechanisms are not implemented. By analysing the attributes of messages exchanged between SCF and SPF, as well as other collected messages, an adversary can *link* relevant data flows and *identify* sensitive information about targets, UE identities, app identity, geo-location, etc. *Repudiation* threats may also arise if communication logging is insufficient. At a lower level, an adversary can perform traffic fingerprinting by observing entity traffic, extracting sensing attributes in a given deployment. Similarly, a number of *tampering* threats could lead to *non-compliance* if policies at SCF/SPF are tampered with.

### 5.3.7    Threats in NEF ⟷ SPF Interface

SPF transmits the result of a sensing session (RESULT_STREAM) to the requesting entity APP via NEF, which ensures the application of correct sensing policies to the result. Numerous threat categories are possible to the communication between NEF and SPF. Unauthorised entities may *impersonate* SPF and transmit manipulated or malicious sensing results to APP. Similarly, unauthorised alterations to RESULT_STREAM could lead to the injection or removal of objects in the sensing result. A *privilege escalation* attack at NEF could bypass policy checks and expose sensing results to unauthorised entities, including NEF, internal entities, or external applications. Failure to apply required privacy controls as per the sensing policy to the RESULT_STREAM may result in *non-compliance* threats within the network. Moreover, an adversary monitoring data on this interface could *identify* or *link* sensitive information about sensing entities, target environments, and sensing activities within the environment.

### 5.3.8    Threats in SPCTM ⟷ SCF/SPF/NEF Interface

Communication between SPCTM and other functions involve sensitive information, such as sensing policies, consent information, and disclosure logs. Therefore, the interfaces that involve SPCTM, can be the target for many threats. Without proper authentication mechanisms and encryption, unauthorised entities may *spoof* SENS_POL_REQ to SPCTM, masquerading as legitimate NEF, to ascertain what an APP/entity is authorised to sense. Additionally, the interface may *disclose* sensitive information, such as sensing policies, the type of sensing requested by an entity/APP, and comprehensive logs of sensing requests and results. Moreover, adversaries could extract sensitive information, such as processing policies corresponding to a subject/APP/area, results of sensing sessions, or valuable data (e.g., IP addresses of operators), leading to *linkability* and *identifiability* threats. *Non-compliance* threats may arise from incorrect functioning or malicious tampering at SPCTM, potentially resulting in the retrieval of incorrect sensing and processing policies in SPF, SCF, and NEF.

## 6    Discussion and Mitigation Strategies

### 6.1    Privacy Enforcement Through SPCTM and SPF

In this work, we proposed SPCTM as the governing entity for the privacy framework in JCAS, supporting mechanisms for transparency, consent management, sensing policy management, and responsible accountability, along with authentication and authorisation. Although SPCTM does not directly enforce privacy controls itself, it serves as a governance framework that oversees various aspects of privacy and regulatory requirements. Transparency data, coupled with the sensing transparency function, informs users about their involvement in sensing activities, complying with GDPR requirements for the right to be informed and ensuring awareness of ongoing sensing activities. Similarly, consent data managed by sensing consent management function ensures that sensing activities respect the consent of affected data subjects, while also adhering to the latest consent policies during sensing and data processing. Sensing policies and current session policies stored at SPCTM collaborate with the sensing policies decision point function, enabling all sensing functions (SCF and SPF) to comply with the most recent set of policies. Additionally, NEF utilises the SPD point to authorise an application or entity for a given sensing request, while also maintaining sensing-related logs at SPCTM for accountability and transparency purposes.

On the other hand, we proposed short-lived sensing processing session in SPF. These ephemeral sessions in SPF are orchestrated by SPF with privacy controls, which are applicable on disclosure as per the sensing policy. These privacy controls ensures that the data being written to the sensing store safeguards the PIIs. Similarly, before sending the sensing result (RESULT_STREAM) to app (through NEF), necessary privacy controls should be applied to the result to disclose only necessary information to app in a secure way. While not solving all privacy concerns, properly configured transient SPF sessions offer a mechanism to address numerous privacy issues.

## 6.2  Suggested Security and Privacy Controls

We suggest some of the essential security and privacy controls for JCAS system. To minimise the overall negative effect on system's functionality, dependability, safety, and to improve cost efficiency, we advocate for the security and privacy solutions that are the most necessary for mitigating the threats discussed in Sect. 5.3.

### 6.2.1  Identification and Authentication

Establishing high levels of assurance for the information originating and communicated by the components, modules, and interfaces in JCAS necessitates ensuring the authenticity of that information. The latter usually use corresponding hardware and software roots of trust inherent to these components, modules, and interfaces [42]. Further steps in establishing authenticity can involve various tools and methods.

Identification and authentication in JCAS, whether hardware or software modules, must establish and verify claimed identities and associated security attributes. Due to the involvement of sensitive entities, such as targets and the SUs and their sensitive information, solutions should validate entity identities, their authority to interact with other entities, and ensure correct association of security attributes [22]. Furthermore, solutions should establish parameters such as the maximum number of unsuccessful authentication attempts to reduce the vulnerability to brute-force attacks. It is also suggested that suitable actions for failures, such as entity lockout or triggering alerts for further investigation, be outlined to enhance system security.

Attention should be paid to authentication mechanisms supported by JCAS components and modules and the properties of the attributes on which they are based. For example, unforgeable authentication prevents the forging or copying of authentication data, while single-use mechanisms operate with data for one-time use [33].

**6.2.2  Data Protection** Ensuring the integrity, confidentiality, and access control of signals, sensed data, and associated security attributes within JCAS modules is paramount. This entails establishing policies for data protection, implementing techniques for data flow protection, and managing offline storage, import, and export procedures.

- **Access and information flow control**: Access control policies define security behaviours enforceable within JCAS, outlining requirements for relevant security techniques and means. This control scope involves three main elements: the subjects and objects governed by the policy and the operations covered by it [5, 23]. Information flow control policies delineate control scope, characterised by three sets: controlled subjects, controlled information, and operations governing information flow to and from controlled subjects [6]. For instance, as proposed in this paper, the SPCTM component (see Fig. 8) incorporates access control policies and information flow control policies.
- **Information retention and disposal**: We recommend implementing information retention control in JCAS to securely manage data that is no longer needed by components or modules. This includes deleting copies of specified objects or data when they are no longer necessary for operation and defining necessary operations for each object [25]. Additionally, residual information protection ensures that data in a resource is not accessible when the resource is de-allocated from one entity and reallocated to another, preventing data leakage. Furthermore, it is necessary to safeguard data stored in a resource that has been logically deleted or released but may persist within the controlled resource, potentially being reallocated to another object [19].
- **Integrity and confidentiality of sensing data-in-store**: To protect the integrity of sensing data and associated PIIs, we suggest implementing rollback solutions, which revert the last operation or a sequence of operations within a defined limit, such as a time period, and restore to a previously known state [44]. Integrity solutions for data stored in the sensing store should safeguard it within a component or module, monitoring and correcting errors that may impact data stored in memory or storage devices [31]. We suggest the confidentiality of sensing data stored in the JCAS components and modules to restrict access to memory data solely through specified interfaces and prevent unauthorised information access. The specifics of these confidentiality solutions may vary based on designated memory areas, cryptographic methods, or the necessity of JCAS stakeholder intervention [15].
- **Integrity and confidentiality of signals-and-data-in-transit**: To ensure the integrity of control signals and sensed data transmitted across JCAS interfaces, we propose employing solutions capable of detecting modifications, deletions, insertions, and replay errors [37]. Additionally, for data recovery at the receiving end, options include utilising source assistance (e.g., automatic repeat query) or standalone recovery methods (e.g., forward error correction) [4]. Moreover, it is imperative to implement confidentiality-based solutions to safeguard JCAS data from disclosure during transit.
- **Security of imported/exported sensing data**: We recommend employing solutions for data authentication, export, and import to secure offline

operations on sensing data. Data authentication allows an entity to verify the authenticity of the information, ensuring the validity of specific data units and preventing forgery or fraudulent modification [29]. Depending on the use case, exporting data from the JCAS component or module (e.g., to a third party) should either maintain security attributes and data protection or discard them post-export. This security feature focuses on export limitations and the association of security attributes with exported user data. Similarly, security techniques for importing data into JCAS (e.g., from third parties) must address import limitations, define desired security attributes, and interpret associated security attributes with the imported data [43].

**6.2.3   Privacy Controls**   Here, we discuss the core privacy controls and extended privacy controls. Core privacy controls safeguard an entity's PII from discovery and misuse, encompassing anonymity, pseudonymity, unlinkability, and unobservability. We also emphasise the need for extended privacy controls, including consent-and-transparency-enabling policies concerning sensing activities. The following details of the core privacy controls shall be considered.

- **Anonymisation and pseudonymisation**: During sensing, techniques such as data aggregation, randomisation, and masking, could obscure identifying information of targets and sensing units during sensing data processing [13]. Different encryption and tokenisation approaches can further safeguard sensitive data while transmitting sensing signals from the SU to the SPF. Further, differential privacy methods add noise to data, preserving statistical properties, while dynamic pseudonymisation assigns temporary identifiers to prevent long-term tracking. Implementing these measures can ensure privacy in sensing systems throughout data processing, transmission, and analysis, allowing for valuable insights while protecting individual privacy [7].
- **Unlinkability and unobservability**: Unlinkability of sensing requests from the applications and responses is essential for upholding privacy rights, for instance, preventing unauthorised tracking [32]. Further, configuration messages from the SCF to the SUs and the responses from the SUs need methods to ensure unlinkability. When applications make sensing requests, unlinkability may be achieved by anonymising the requester's identity through techniques like dynamic pseudonymisation or session IDs. During sensing activities, unlinkability may be achieved by obfuscation methods to introduce randomness into data, data fragmentation for making the reconstruction process challenging, and employing cryptographic or physical layer security methods to protect data transmission. This ensures that data collected from targets cannot be easily linked back to specific individuals or entities. Unobservability of sensing units and network resources involved in sensing may be achieved through covert operations and channels, noise injection, anonymisation, data fragmentation, decoy traffic, mix networks [30], and differential privacy. These methods, whether used individually or in combination, aim to conceal and confuse sensitive sensing activities or data, thereby safeguarding privacy.

- **Extended privacy controls**: Privacy solutions for "consent and choice" ensure appropriate handling of PIIs in JCAS, specifying methods, timing, and conditions for processing within corresponding modules and components [16]. For instance, consent initiation may occur within the SPCTM component as depicted in Fig. 1. To comply with the principle of "openness, transparency, and notice" in sensing, stakeholders should have access to general information regarding the handling of PII policy. JCAS components and modules must implement appropriate solutions to inform relevant stakeholders about any changes to the policy [24]. In our proposal, SPCTM is considered for the above functions.

**6.2.4    Security and Privacy Controls for UE** When user equipment integrates sensing units, it is crucial to enforce security and privacy controls on the UE to safeguard sensitive sensing data and preserve the UE's interests in consent and policy management. Functionalities similar to SPCTM on the UE side are necessary to manage sensing policies, obtain consent to enable UE sensing services and ensure transparent data disclosure to the network. Additionally, due to the involvement of UE resources in processing sensing data, network functions similar to SCF and SPF on the network side are necessary for sensing management in UE. Figure 9 illustrates our envisioned interfaces and functions on the UE side concerning JCAS.

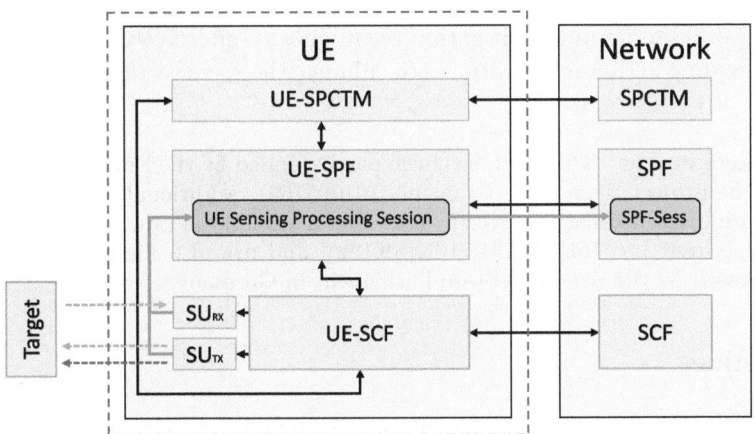

**Fig. 9.** Illustration of UE specific controls and processing for sensing.

Further, the controls on UE have to include data encryption during transmission and storage, strict access control mechanisms for enabling sensing services in the UE, and authentication requirements. Anonymisation and pseudonymisation techniques have to safeguard the PII and sensing information. With proper consent management, UEs can gain control over the sensing units, and logging will allow maintaining records of the sensing activities [18]. Implementing secure communication protocols and conducting regular privacy impact assessments can further enhance security and privacy measures, ensuring compliance with relevant regulations and standards.

# 7   Conclusions and Future Work

In this paper, we examined the architectural, security, and privacy aspects of JCAS, a vital technology anticipated in 6G networks. Drawing from our analysis on the emergent JCAS architecture, we proposed some enhancements involving additional network functions and interfaces to address key challenges related to SPCTM and privacy of PIIs involved in sensing. Subsequently, we performed a detailed threat analysis for each interface within the proposed JCAS architecture, aligning with standard CWEs and presenting a threat summary table synchronised with JCAS threats. To mitigate the security and privacy risks associated with JCAS, we put forth security and privacy controls, emphasising their significance for JCAS systems.

As part of the future extension of this work, we intend to conduct a more comprehensive risk assessment of the proposed JCAS architecture, including a detailed analysis of threat likelihood and potential impact. We will explore how threats to sensing activities affect the communication systems. Additionally, we plan to propose fine-grained privacy controls within the JCAS framework to ensure stronger protection of user data and privacy rights. Furthermore, we intend to develop detailed mitigation techniques to effectively address identified risks, providing actionable strategies to enhance the security and privacy of JCAS systems.

**Acknowledgement.** This work has been partly funded by the European Commission through the project Hexa-X-II (Grant no. 101095759). Additionally, the authors from Barkhausen Institute are supported by the Federal Ministry of Education and Research, Germany (Grant no. 16KISK231, 16KISK122), and are also financed based on the budget passed by the Saxonian State Parliament in Germany.

# References

1. 3GPP: System architecture for the 5G System (5GS) (Release 17). Technical report, 3GPP TS 23.501 V 17.7.0, 3GPP (2023)
2. 3GPP: Feasibility Study on Integrated Sensing and Communication (Release 19). Technical report, TR 22.837 V19.2.1, 3GPP (2024)
3. 3GPP: Procedures for the 5G System (5GS); Stage 2 (Release 18). Technical report, 23.502, V 18.5.0, 3GPP (2024)
4. Ahmed, A., Al-Dweik, A., Iraqi, Y., Mukhtar, H., Naeem, M., Hossain, E.: Hybrid automatic repeat request (HARQ) in wireless communications systems and standards: a contemporary survey. IEEE Commun. Surv. Tutor. **23**(4), 2711–2752 (2021)

5. Bonatti, P., De Capitani di Vimercati, S., Samarati, P.: An algebra for composing access control policies. ACM Trans. Inf. Syst. Secur. **5**(1), 1-35 (2002)
6. Broberg, N., Sands, D.: Flow-sensitive semantics for dynamic information flow policies. In: Proceedings of the ACM SIGPLAN Fourth Workshop on Programming Languages and Analysis for Security, PLAS 2009, pp. 101–112. Association for Computing Machinery, New York (2009)
7. Carvalho, T., Moniz, N., Faria, P., Antunes, L.: Survey on privacy-preserving techniques for data publishing. arXiv preprint arXiv:2201.08120 (2022)
8. Christin, D., Reinhardt, A., Kanhere, S.S., Hollick, M.: A survey on privacy in mobile participatory sensing applications. J. Syst. Softw. **84**(11), 1928–1946 (2011)
9. Dass, P., Ujjwal, S., Novotny, J., Zolotavkin, Y., Laaroussi, Z., Köpsell, S.: Addressing privacy concerns in joint communication and sensing for 6G networks: challenges and prospects. arXiv preprint arXiv:2405.01742v2 (2024). http://arxiv.org/abs/2405.01742v2
10. De Cristofaro, E., Soriente, C.: Participatory privacy: enabling privacy in participatory sensing. IEEE Network **27**(1), 32–36 (2013)
11. Deng, M., Wuyts, K., Scandariato, R., Preneel, B., Joosen, W.: A privacy threat analysis framework: supporting the elicitation and fulfillment of privacy requirements. Requirements Eng. **16**(1), 3–32 (2011)
12. Gisdakis, S., Giannetsos, T., Papadimitratos, P.: Sppear: security & privacy-preserving architecture for participatory-sensing applications. In: Proceedings of the 2014 ACM Conference on Security and Privacy in Wireless & Mobile Networks, pp. 39–50 (2014)
13. Gjermundrød, H., Dionysiou, I., Costa, K.: privacyTracker: a privacy-by-design GDPR-compliant framework with verifiable data traceability controls. In: Casteleyn, S., Dolog, P., Pautasso, C. (eds.) ICWE 2016. LNCS, vol. 9881, pp. 3–15. Springer, Cham (2016). https://doi.org/10.1007/978-3-319-46963-8_1
14. Günlü, O., Bloch, M., Schaefer, R.F., Yener, A.: Secure joint communication and sensing. In: 2022 IEEE International Symposium on Information Theory (ISIT), pp. 844–849. IEEE (2022)
15. Henson, M., Taylor, S.: Memory encryption: a survey of existing techniques. ACM Comput. Surv. **46**(4) (2014)
16. Ho, J.T., Dearman, D., Truong, K.N.: Improving users' security choices on home wireless networks. In: Proceedings of the Sixth Symposium on Usable Privacy and Security, SOUPS 2010. Association for Computing Machinery, New York (2010)
17. Howard, M., Lipner, S.: The Security Development Lifecycle, vol. 8. Microsoft Press Redmond (2006)
18. ISO/IEC: Information technology - security techniques - privacy architecture framework. Technical report, ISO/IEC 29101, 2018-11, ISO/IEC (2018)
19. Jordon, M.: Cleaning up dirty disks in the cloud. Netw. Secur. **2012**(10), 12–15 (2012)
20. Kapoor, P., Vora, A., Kang, K.D.: Detecting and mitigating spoofing attack against an automotive radar. In: 2018 IEEE 88th Vehicular Technology Conference (VTC-Fall), pp. 1–6. IEEE (2018)
21. Klasnja, P., Consolvo, S., Choudhury, T., Beckwith, R., Hightower, J.: Exploring privacy concerns about personal sensing. In: Tokuda, H., Beigl, M., Friday, A., Brush, A.J.B., Tobe, Y. (eds.) Pervasive 2009. LNCS, vol. 5538, pp. 176–183. Springer, Heidelberg (2009). https://doi.org/10.1007/978-3-642-01516-8_13
22. Klingenstein, N.: Attribute aggregation and federated identity. In: 2007 International Symposium on Applications and the Internet Workshops, pp. 26–26 (2007)

23. Kolovski, V., Hendler, J., Parsia, B.: Analyzing web access control policies. In: Proceedings of the 16th International Conference on World Wide Web, WWW 2007, pp. 677–686. Association for Computing Machinery, New York (2007)

24. Laoutaris, N.: Data transparency: Concerns and prospects [point of view]. Proc. IEEE **106**(11), 1867–1871 (2018)

25. Li, J., Singhal, S., Swaminathan, R., Karp, A.H.: Managing data retention policies at scale. IEEE Trans. Netw. Serv. Manage. **9**(4), 393–406 (2012)

26. Liu, B., et al.: Architecture for cellular enabled integrated communication and sensing services. China Commun. **20**(9), 59–77 (2023)

27. Martins, Ó., Vilela, J.P., Gomes, M.: Poster: Privacy-preserving joint communication and sensing. In: 2023 IEEE 24th International Symposium on a World of Wireless, Mobile and Multimedia Networks (WoWMoM), pp. 329–331. IEEE (2023)

28. MITRE: CWE version 4.14. https://cwe.mitre.org/data/downloads.html. Accessed 4 Apr 2024

29. Nizamuddin, N., Hasan, H.R., Salah, K.: IPFS-blockchain-based authenticity of online publications. In: Chen, S., Wang, H., Zhang, L.-J. (eds.) ICBC 2018. LNCS, vol. 10974, pp. 199–212. Springer, Cham (2018). https://doi.org/10.1007/978-3-319-94478-4_14

30. Sampigethaya, K., Poovendran, R.: A survey on mix networks and their secure applications. Proc. IEEE **94**(12), 2142–2181 (2006)

31. Sivathanu, G., Wright, C.P., Zadok, E.: Ensuring data integrity in storage: techniques and applications. In: Proceedings of the 2005 ACM Workshop on Storage Security and Survivability, StorageSS 2005, pp. 26–36. Association for Computing Machinery, New York (2005)

32. Steinbrecher, S., Köpsell, S.: Modelling unlinkability. In: Dingledine, R. (ed.) PET 2003. LNCS, vol. 2760, pp. 32–47. Springer, Heidelberg (2003). https://doi.org/10.1007/978-3-540-40956-4_3

33. Su, J., Cao, D., Zhao, B., Wang, X., You, I.: ePASS: an expressive attribute-based signature scheme with privacy and an unforgeability guarantee for the internet of things. Future Gener. Comput. Syst. **33**, 11–18 (2014). Special Section on Applications of Intelligent Data and Knowledge Processing Technologies; Guest Editor: Dominik Ślęzak

34. Sun, Z., Balakrishnan, S., Su, L., Bhuyan, A., Wang, P., Qiao, C.: Who is in control? practical physical layer attack and defense for mmWave-based sensing in autonomous vehicles. IEEE Trans. Inf. Forensics Secur. **16**, 3199–3214 (2021)

35. Wang, X., Mei, J., Cui, S., Wang, C.X., Shen, X.S.: Realizing 6G: the operational goals, enabling technologies of future networks, and value-oriented intelligent multidimensional multiple access. IEEE Network **37**(1), 10–17 (2023)

36. Wei, Z., Liu, F., Masouros, C., Su, N., Petropulu, A.P.: Toward multi-functional 6G wireless networks: integrating sensing, communication, and security. IEEE Commun. Mag. **60**(4), 65–71 (2022)

37. Wicker, S.B.: Error Control Systems for Digital Communication and Storage, vol. 1. Prentice Hall, Englewood Cliffs (1995)

38. Wild, T., Braun, V., Viswanathan, H.: Joint design of communication and sensing for beyond 5G and 6G systems. IEEE Access **9**, 30845–30857 (2021)

39. Wild, T., Grudnitsky, A., Mandelli, S., Henninger, M., Guan, J., Schaich, F.: 6G integrated sensing and communication: from vision to realization. In: 2023 20th European Radar Conference (EuRAD), pp. 355–358. IEEE (2023)

40. Wymeersch, H., et al.: Joint communication and sensing for 6G–a cross-layer perspective. arXiv preprint arXiv:2402.09120 (2024)

41. Wymeersch, H., et al.: Deliverable d3.3 final models and measurements for locali-
    sation and sensing. Technical report, Hexa-X (2023)
42. Yang, T.W., Ho, Y.H., Chou, C.F.: Achieving M2M-device authentication through
    heterogeneous information bound with USIM card. Futur. Gener. Comput. Syst.
    **110**, 629–637 (2020)
43. Yang, Z., Tang, J., Liu, H.: Cloud information retrieval: model description and
    scheme design. IEEE Access **6**, 15420–15430 (2018)
44. Zhou, W., Yuan, S., Li, L., Yeh, K.H.: A novel fast recovery method for HT tamper
    in embedded processor. In: Meng, W., Li, W. (eds.) BlockTEA 2022, pp. 131–139.
    Springer, Cham (2023). https://doi.org/10.1007/978-3-031-31420-9_8
45. Zhu, P., Ma, J., Bayesteh, A., Chen, Y., Tong, W.: Integrated sensing and com-
    munication network. US Patent App. 18/324,458 (2023)

# The Lawfulness of Re-identification Under Data Protection Law

Teodora Curelariu[1]($\boxtimes$) and Alexandre Lodie[2]

[1] CESICE, Université Grenoble-Alpes, Centre Inria de l'Université Grenoble-Alpes,
Montbonnot-Saint-Martin, France
teodora.curelariu@inria.fr
[2] Centre Inria de l'Université Grenoble-Alpes, Montbonnot-Saint-Martin, France

**Abstract.** Data re-identification methods are becoming increasingly sophisticated and can lead to disastrous data breaches. Re-identification is a key research topic for computer scientists as it can be used to reveal vulnerabilities of de-identification methods such as anonymisation or pseudonymisation. However, re-identification, even for research purposes, involves processing personal data. From this background, this paper aims to investigate whether re-identification carried out by computer scientists for research purposes can be considered GDPR-compliant. This issue is paramount to contribute to improving the state of knowledge concerning data security measures.

**Keywords:** Re-identification · Computer Science · PETs · Personal Data · GDPR · Data Protection

## 1 Introduction

Data are being increasingly shared on a wide scale, be it for public re-use[1] but also for marketing purposes.[2,3] Nonetheless, this trend comes with significant privacy concerns[4], particularly regarding the re-identification of individuals through data mining[5].

This poses risks such as unauthorised access, misuse of personal information and disclosure of personal data, ultimately undermining de-identification techniques and privacy safeguards.

---

[1] Peloquin, D., DiMaio, M., Bierer, B. et al. Disruptive and avoidable: GDPR challenges to secondary research uses of data. Eur J Hum Genet 28, 697–705 (2020).

[2] Sheth, J., Charles H., Next Frontiers of Research in Data Driven Marketing: Will Technique Keep up with Data Tsunami?. Journal of Business Research, vol. 125, 780–84 (2021).

[3] Rogers, J., Song A., Digital Marketing in The Legal Profession: What's Going On and Does It Matter? Law, Technology and Humans, vol. 5, no. 2, 134 – 64 (2023).

[4] Henriksen-Bulmer, J., Jeary, S.: Re-identification attacks—A systematic literature re-view, International Journal of Information Management, 36, 1184–1192 (2016).

[5] Schermer, B.W., The Limits of Privacy in Automated Profiling and Data Mining. Computer Law & Security Review, (2011).

M. Jensen et al. (Eds.): APF 2024, LNCS 14831, pp. 112–131, 2024.
https://doi.org/10.1007/978-3-031-68024-3_6

De-identification, as defined by the NIST,[6] serves as a mechanism for organisations "to remove personal information from data that they collect, use, archive, and share with other organisations.[7]" As for re-identification, it can be defined as "a process by which information is attributed to de-identified data in order to identify the individual to whom the de-identified data relate.[8]"

Re-identification of de-identified data challenges the effectiveness of data protection techniques, which are designed to safeguard personal data by preventing the direct or indirect identification of data subjects. From this standpoint, re-identification is perceived as a security risk that must be dealt with by computer scientists and legal practitioners. The very definition of personal data underscores that "personal data are any information which are related to an identified or identifiable natural person.[9]" It is worth mentioning that according to these definitions, identifiability is the core criterion to define personal data. The importance of identifiability has been underlined by the literature as well.[10] Despite its importance, the definition of identifiability remains uncertain. Indeed, this is a contentious issue which is widely debated by scholars, some of them advocating for an objective approach of the identifiability of data subjects, while other stand for a more relative approach.[11] As regards the former approach, the qualification of data as personal data depends on the inherent features of the data themselves.[12] The focus is on whether data alone permit the re-identification of data subjects, no matter who holds them. When it comes to the latter approach, identifiability is more context-related and depends on the means and additional information in the hands of the person or organisation holding data.[13]

Thus, the choice of approach can directly influence the way data protection law applies. When the objective approach prevails, data may be considered personal if there is an abstract possibility to re-identify them. This broad interpretation places a greater burden on data controllers to ensure compliance with privacy regulations and to implement robust measures to personal data.

*A contrario*, in jurisdictions favouring the relative approach, the determination of whether data qualifies as personal may depend on the specific circumstances surrounding its processing and the likelihood of re-identification. This approach offers more flexibility, for instance in matter of data re-use for research purposes,[14] but may also

---

[6] National Institute of Standards and Technology.

[7] NIST, https://www.nist.gov/itl/iad/deidentificationnistgov, last accessed 2024/02/22.

[8] NIST, https://csrc.nist.gov/glossary/term/re_identification, last accessed 2024/02/22.

[9] See Article 4 of the GDPR.

[10] Spindler, G., Schmechel, P.: Personal Data and Encryption in the European General Data Protection Regulation., JIPITEC, 164 (2016).

[11] Zuiderveen Borgesius, F., The Breyer Case of the Court of Justice of the European Union: IP Addresses and the Personal Data Definition., European Data Protection Law Review, Vol 3, Issue 1, 130–137 (2017).

[12] OPINION OF ADVOCATE GENERAL CAMPOS SÁNCHEZ-BORDONA delivered on 12 May 2016, Case C-582/14 Patrick Breyer v Bundesrepublik Deutschland, § 52.

[13] Ibid., § 53.

[14] Mourby, M. et al.: Are 'pseudonymised' data always personal data? Implications of the GDP for administrative data research in the UK., Computer Law & Security Review, Vol 34, 222 – 233, (2018).

create uncertainty and inconsistency in legal interpretations.[15] To sum up, the identifiability criterion involves assessing the likelihood of re-identification of data subjects,[16] and thus the applicability of data protection regulations.

To assess the likelihood of re-identification, a data controller must take into account some objective factors such as "the available technology at the time of the processing and technological developments.[17]". Put differently, the development of re-identification techniques and technology has an impact on the assessment of the robustness of a de-identification scheme. It contributes to what is considered the "state-of-the-art" in matter of data security.[18] The more sophisticated re-identification is, the stronger de-identification will be. Once de-identification is considered to be achieved and involves only a residual risk of re-identification, data are said to be anonymised.[19]

Quite surprisingly, re-identification is not considered as an independent research area, at least for legal scholars. It is mainly seen as a means to assess the reliability of data protection techniques and whether data have been properly anonymised with regard to the requirements of the GDPR.[20]

De-identification involves various data protection methods including anonymisation, but also pseudonymisation. The latter is mentioned as a security measure which can help protect data under Article 32 of the GDPR. This article underlines that data must be granted an appropriate level of security, taking into account the "state-of-the-art". These elements show that there is a pressing need for computer science research on this field. No matter if data are said to be anonymised or only pseudonymised, they can be re-identified so that re-identification is a risk to be taken into account when implementing data protection techniques.

As data controllers have obligations regarding the implementation of cutting-edge security measures, research is needed in order to keep de-identification techniques up-to-date. However, the lawfulness of re-identification under EU data protection law remains uncertain, as it inherently involves bypassing data protection measures.

This paper aims to investigate why carrying out research in the field of re-identification is key to improve privacy. It also intends to assess whether re-identification techniques implemented for research purposes can be considered GDPR compliant.

In the following section, technical aspects related to re-identification are discussed, and how research in this area contributes to enhancing privacy. Building on this technical foundation, Sect. 3 will question the lawfulness of re-identification under EU data

---

[15] Lodie A., Case C-479/22 P, Case C-604/22 and the limitation of the relative approach of the definition of 'personal data' by the ECJ., European Law Blog, (2024).

[16] See recital 26 of the GDPR.

[17] Ibid.

[18] Esayas S. Y., The role of anonymisation and pseudonymisation under the EU data privacy rules: beyond the 'all or nothing' approach., European Journal of Law and Technology, Vol 6, No 2, 19, (2015).

[19] Finck, M., Pallas, F.: They who must not be identified—distinguishing personal from non-personal data under the GDPR., International Data Privacy Law, 2020, Vol. 10, No. 1, 35 (2020).

[20] Stalla-Bourdillon, S., Knight, A.: Anonymous Data V. Personal Data—A False Debate: An EU Perspective on Anonymization, Pseudonymization and Personal Data., International Data Privacy Law, Vol. 10, No. 1., (2020).

protection law. In the final section, some guidance is provided to computer scientists to help them re-identify data in a way compliant with the GDPR. Some contentious points will be underlined.

## 2   Technical Aspects Related to Re-identification

In order to understand why re-identification is paramount to improve security measures implemented on personal data, a quick background on re-identification is needed.

Over the years, re-identification attacks have become increasingly sophisticated and effective. In the current landscape, any de-identified data can potentially be subject to re-identification.[21] This is primarily due to the widespread availability and accessibility of data sources and online datasets that contain vast amounts of personal data. As previously mentioned, it is important to note that re-identification attacks have capitalised on progress in machine learning and other AI applications,[22] so that their potential and effectiveness cannot be overstated. Data controllers must be prepared to address these threats and the resulting consequences, including data breaches and risks to individuals' privacy.

Re-identification attacks have been successful on various kinds of data, such as health data, movie preferences, location data, university courses and users' search queries (see Table 1). Other techniques have been deployed to enable the free use of data without privacy risks, including synthetic data, but even in the latter case, data can be re-identified.[23] The nature of data is therefore irrelevant to study re-identification since no de-identified data is immune from re-identification attacks.

The increasing ease of cross-referencing data has changed how re-identification attacks are carried out. Cross-referencing refers to the process of comparing information from one dataset with information from another dataset: this information within these datasets is compared to identify matches.[24] This comparison can be based on specific identifiers or attributes. Datasets containing personal information often include various elements alongside the actual data, such as direct identifiers (e.g., social security numbers), indirect identifiers (e.g.; ZIP codes), and quasi-identifiers (attributes such as ZIP code, birthdate and gender).[25] While a direct identifier permits to uniquely identify an individual without additional knowledge, an indirect identifier permits such identification when combined with additional information. A quasi-identifier cannot uniquely identify an individual, but it is sufficiently well correlated with the individual. A set of quasi-identifiers can constitute a profile which uniquely identifies the individual. By cross-referencing a de-identified dataset with other sources, an adversary can obtain a set

---

[21] Ibid.

[22] Rocher, L. et al.: Estimating the Success of Re-Identifications in Incomplete Datasets Using Generative Models., Nature communications, (2019).

[23] Giomi, M. et al., A Unified Framework for Quantifying Privacy Risk in Synthetic Data., (2022).

[24] Yang, H., Yi, D., Liao, S., Lei, Z., & Li, S., Cross Dataset Person Re-identification. In ACCV Workshop., (2015).

[25] Garfinkel, S., De-Identification of Personal Information, NIST Interagency/Internal Report (NISTIR), National Institute of Standards and Technology, Gaithersburg, MD, (2015).

of quasi-identifiers that uniquely identifies an individual, potentially revealing her/his true identity/name.

The increasing availability of data sources for cross-referencing purposes has enabled anyone to use online and freely available data (for example, from social networks[26] such as IMDB or LinkedIn) to re-identify individuals in large datasets. Data openness and cross-referencing techniques both emphasise the following statement of the NIST: "[...]it is not possible to algorithmically determine what kinds of contextual information can be used to assist in future re-identification efforts.[27]" In other words, it is difficult to predict the capabilities of an adversary to undertake a re-identification attack in the future.

When a re-identification attack occurs, three types of information can be disclosed:[28] identity, attribute and inferred information. Identity disclosure occurs when an attacker successfully links de-identified data to a specific individual, directly revealing their identity. Attribute disclosure occurs when the adversary can attribute a piece of information to an individual without necessarily knowing their identity, thereby revealing personal attributes associated with that individual. Inferential disclosure occurs, according to the NIST, "when information can be inferred with high confidence from statistical properties of the released data[29]" providing insights into sensitive details about individuals even without direct identification. It is crucial to note that re-identification encompasses more than just identity disclosure, and all forms of disclosure, including attribute and inferential, must be carefully considered when assessing the risks associated with re-identification attacks, as they can lead to the processing of personal data.

Re-identification techniques grow more sophisticated alongside advancements in algorithms and availability of diverse data sources for cross-referencing. This risk extends to various forms of data, such as statistics, aggregated data,[30] and even machine learning models,[31] where individuals might be linked back to the initial data.

Furthermore, re-identification attacks also benefit from advancements in artificial intelligence. Models can be specifically trained to re-identify individuals, by singling them out or by performing inference attacks, exploiting datasets. Attacks based on machine learning have been carried out on medical data[32] or connection data.[33]

We can also speculate that quantum computers will also enhance the possibilities and effectiveness of re-identification attacks due to their ability to break traditional

---

[26] De Montjoye Y-A., Hidalgo, C. A., Verleysen, M., Blondel, V. D., Unique in the Crowd: The Privacy Bounds of Human Mobility. Scientific reports, 3, 1376, (2013).

[27] Garfinkel, S., De-Identification of Personal Information, NIST Interagency/Internal Report (NISTIR), National Institute of Standards and Technology, Gaithersburg, MD, (2015).

[28] Ibid.

[29] Ibid.

[30] Willemson, J. (2022). Fifty Shades of Personal Data – Partial Re-identification and GDPR. In: Gryszczyńska, A., Polański, P., Gruschka, N., Rannenberg, K., Adamczyk, M. (eds) Privacy Technologies and Policy. APF 2022. Lecture Notes in Computer Science, vol 13279. Springer, Cham (2022).

[31] Shokri, R.: Membership Inference Attacks Against Machine Learning Models. (2017).

[32] Rocher, L. et al.: Estimating the Success of Re-Identifications in Incomplete Datasets Using Generative Models, Nature communications, 10, 3069 (2019).

[33] De Montjoye, Y.-A., et al,.: Unique in the Crowd: The Privacy Bounds of Human Mobility. 3, 1376 (2013).

encryption methods through advanced algorithms.[34]The main technology used to secure data is cryptography (and particularly encryption). Cryptography is a fundamental aspect of privacy-enhancing technologies (PET), and the evolution of quantum computers may further augment the capabilities and efficacy of re-identification attacks, potentially presenting new challenges to data privacy and security.

The following table aims to give some insights into the most emblematic re-identification attacks.

**Table 1.** The rise of re-identification attacks.

| Year | Dataset creator | Type of data | Attack type | References |
|---|---|---|---|---|
| 1997 | NAHDO GIC Cambridge Massachusetts | Hospitalisation records Medical records in the GIC data Voter registration data | Crossing databases | [26] |
| 2000 | NAHDO Cambridge Massachusetts | Hospitalisation records Voter registration data | Crossing databases (census) | [27] |
| 2006 | AOL | Users' search queries | Crossing databases | [31] |
| 2007 | Netflix | Users' movie preferences | Crossing databases | [17] |
| 2008 | Cabspotting | Taxi trajectory (GPS coordinates) | Point of interests discovery | [6] |
| 2017–2018 | Strava | Users' trajectories (GPS coordinates) | Regression | [3] |
| 2021 | edX (Harvard) | Students enrolled in edX courses | Crossing databases | [1] |

The table shows how re-identification attacks are exploiting additional data obtained by different means. Early examples include Latanya Sweeney re-identifying a governor using voter records and hospital data.[35]More recently, companies like AOL[36] and Netflix[37] released search history and movies ratings, that, when combined with other

---

[34] European Union Agency for Network and Information Security, https://www.enisa.europa.eu/publications/privacy-and-data-protection-by-design, last accessed 2024/06/12.

[35] Sweeney, L.: K-ANONYMITY: A MODEL FOR PROTECTING PRIVACY.., International Journal on Uncertainty, Fuzziness and Knowledge-based Systems 10 (5), 557–570 (2002).

[36] TechCrunch, https://techcrunch.com/2006/08/06/aol-proudly-releases-massive-amounts-of-user-search-data/, last accessed 2024/04/29.

[37] Wired, https://www.wired.com/images_blogs/threatlevel/2009/12/doe-v-netflix.pdf, last accessed 2023/01/25.

information, enabled people to be re-identified. Furthermore, the EdX incident exemplifies the evolving threat of re-identification attacks. Even with supposedly advanced de-identification methods, researchers were still able to re-identify users. These cases highlight the fact that using de-identification techniques and removing identifying information in order to protect data might not be enough to ensure data protection.[38]

Furthermore, most of the re-identification attacks cited in Table 1 have been performed by academics. Public research has a leading role in making progress in re-identification.

Additionally, in the realm of data security, parallels can be drawn between re-identification/anonymisation and cryptanalysis/cryptography. Just as cryptanalysis challenges cryptographic methods to enhance security protocols, re-identification efforts test anonymisation techniques to improve data privacy. Both domains benefit from this ongoing tension: vulnerabilities exposed through re-identification or cryptanalysis lead to stronger anonymisation and cryptographic methods. For instance, early cryptographic methods relied heavily on simple ciphers, which only shifted letters by a fixed number of positions. While these early methods provided basic encryption, they were relatively easy to break with simple frequency analysis.

Over time, as cryptanalysts uncovered these weaknesses, more advanced encryption techniques were developed, which use complex mathematical algorithms. It is important to note that while encryption is not an anonymisation technique, it can serve as a powerful pseudonymisation tool.[39] Re-identification thus remains a potential risk if the encryption key is compromised.

As for re-identification, early anonymisation techniques were only removing direct identifiers in datasets (see the AOL and Netflix examples). Re-identification has proven that it was not enough to prevent an adversary from re-identifying the individuals from those weakly anonymised datasets. Just as cryptanalysis has demonstrated that basic cryptographic methods are insufficient for protecting sensitive data, re-identification has shown the inadequacy of early anonymisation techniques. Both fields thrive on this continuous push and pull, as the advancements in one drive improvements in the other. Nonetheless, more efforts were needed to reach a stronger form of anonymisation. Re-identification is thus still needed today to evaluate new anonymisation proposals.[40]

State-of-the-art re-identification techniques present a tricky dilemma for anonymisation efforts. Enhancing anonymisation also means enhancing re-identification, which may seem paradoxical and counter-intuitive. Without effective anonymisation, individuals' privacy is compromised. However, robust re-identification techniques are also necessary to prevent re-identification attacks on anonymised data. Understanding the legal framework becomes essential to navigate this delicate balance and to preserve the privacy of personal data, within the confines of existing regulations.

---

[38] Mitchum, R., New Kind of Attack Called "Downcoding" Demonstrates Flaws in Anonymizing Data. (2022).

[39] European Data Protection Supervisor, https://www.edps.europa.eu/system/files/2021-04/21-04-27_aepd-edps_anonymisation_en_5.pdf, last accessed 2024/06/13.

[40] Kikuchi H., et al. Ice and Fire: Quantifying the Risk of Re-identification and Utility in Data Anonymization, 2016 IEEE 30th International Conference on Advanced Information Networking and Applications (AINA), Crans-Montana, Switzerland, 1035–1042 (2016).

# 3 What EU Law Says About Re-identification

As a first step, it is worth emphasising what data protection law says about re-identification.

## 3.1 Prohibition of Re-identification Under the Data Governance Act

While the GDPR does not explicitly prohibit re-identification, it emphasises data protection principles that implicitly discourage practices leading to re-identification. However, other legal frameworks in the EU do prohibit re-identification. re-identification.

For instance, the newly adopted Data governance act provides that "[r]e-identification of data subjects from anonymised datasets should be prohibited,[41]" which suggests that the situation regarding the GDPR's treatment of re-identification could change in the future, but, at the moment, it is not explicitly addressed within the GDPR itself.

This same regulation also provides that "[r]e-users shall be prohibited from re-identifying any data subject to whom the data relates and shall take technical and operational measures to pre-vent re-identification and to notify any data breach resulting in the re-identification of the data subjects concerned to the public sector body.[42]" In the latter provision, the prohibition of re-identification seems to be more strictly defined. However, such a conclusion must not be overestimated, as it is mainly designed to frame the situation where a public body shares data with another party, following certain procedures. In this scenario, it is obvious that one of the conditions for sharing data is that the recipient of data will not try to re-identify them. This provision does not address a classic re-identification scheme, where poorly anonymised data are published online and later re-identified by an organisation, or a natural person.

It is worth mentioning that the DGA covers both personal and non-personal data, with the GDPR applying whenever personal data is involved.[43] The regulation emphasises the need for data interoperability and protection against re-identification while encouraging the development of secure data processing environments and standardised anonymisation techniques, which will likely enhance the ability to share data safely and reduce the risk of re-identification. It prohibits re-identification of data subjects from anonymised datasets. However, it is essential to notice that the DGA is not a data protection regulation but rather a legal framework designed to promote data sharing and reuse within the EU.

## 3.2 Identifiability Under EU Data Protection Law

Identifiability and re-identification, though conceptually distinct, often yield similar practical outcomes concerning privacy risks. Both identifiability and re-identification pose significant privacy risks, as they can lead to the exposure of personal information.

---

[41] See recital 8 of the REGULATION (EU) 2022/868 OF THE EUROPEAN PARLIAMENT AND OF THE COUNCIL of 30 May 2022 on European data governance and amending Regulation (EU) 2018/1724 (Data Governance Act).

[42] See Article 5 (5) of the Data Governance Act.

[43] European Commission, https://digital-strategy.ec.europa.eu/en/policies/data-governance-act-explained, last accessed 2024/06/12.

On the one hand, under the GDPR, if data can be linked to an individual, it is considered personal data. The focus is on whether a person can be identified, directly or indirectly, from the data in question. On the other hand, re-identification involves transforming de-identified data back into identifiable data. Thus, while identifiability primarily addresses the inherent link between data and individuals, re-identification underscored the vulnerability of supposedly protected datasets, by demonstrating the potential for originally de-identified data to be transformed into identifiable information. The re-identification process directly influences the identifiability of data.

The European Court of Justice (ECJ) gives us some clues into the issue of re-identification and identifiability under EU data protection law. Indeed, in *Breyer*, the Court assesses whether an IP address can be regarded as personal data for a web service provider. The Court therefore had to evaluate whether the said web service provider had "reasonable means" to identify data, as provided for by Recital 26. While the terminology may differ, the evaluation of identifiability addresses similar concerns as re-identification, as both have the same effects.

From this background the Court claims that the means would not be reasonably likely to be used if "the identification of the data subject was prohibited by law.[44]" This criterion has been recalled by the General Court in case T 557/20 which involved two EU organs and institutions, namely the Single Resolution Board and the EDPS.[45] Essentially, this suggests that if re-identification is made unlawful, said re-identification cannot be deemed reasonable and that the assessment of the means reasonably likely to be used becomes irrelevant. The prohibition of re-identification is thus considered as a silver bullet, allegedly representing the most effective means to protect personal data. It signifies that re-identification cannot be considered reasonable when it contravenes legal restrictions. Due to the absence of a general prohibition of re-identification under the GDPR, the question shifts to whether such prohibitions exist at a national level.

### 3.3 Approaches to Re-identification

Some states have implemented specific regulations and guidelines to address re-identification within their jurisdictions. For instance, the UK has expressly incorporated a provision in the Data Protection Act that prohibits re-identification attacks. Indeed, in a chapter dedicated to "offences relating to personal data", Sect. 171 provides that "[i]t is an offence for a person knowingly or recklessly to re-identify information that is de-identified personal data without the consent of the controller responsible for de-identifying the personal data.[46]" Interestingly, UK law prohibits re-identification as a process, without putting the emphasis on the means by which such re-identification occurs. This provision sets some exceptions to this general rule such as re-identification carried out for public interest reasons, in particular for research purposes. In the explanatory notes of the bill, UK lawmakers underline that this provision tackles the issue of

---

[44] ECJ, JUDGMENT OF THE COURT (Second Chamber), in case C-582/14, Patrick Breyer v. Bundesrepublik Deutschland, 19 October 2016, § 46.

[45] CJEU, JUDGMENT OF THE GENERAL COURT (Eighth Chamber, Extended Composition) in Case T-557/20, Single Resolution Board (SRB) v. European Data Protection Supervisor (EDPS), 26 April 2023.

[46] UK Public General Acts, 2018 c.12, Data Protection Act 2018, legislation.gov.uk.

de-identified data published online, in particular when they are health data which can lead to the re-identification of patients.[47] It must be emphasised here that there have been huge controversies in the UK with regard the re-identification of doctors who have carried out late-term abortion from statistics released by the department of health.[48] Judicial authorities upheld that statistics were not personal data.[49]

Although the GDPR no longer applies in the UK, the principles and legal interpretations remain relevant for understanding how member states might handle re-identification prohibition and how they might evaluate the lawfulness of re-identification means and methods.

### 3.4 Evaluating the Possibility of Criminalising Re-identification

With regard to EU law, recital 149 of the GDPR enables Member states to "lay down the rules on criminal penalties for infringements of this Regulation, including for infringements of national rules adopted pursuant to and within the limits of this Regulation.[50]" One should thus conclude from this provision that it is up to member states to criminalise (or not) re-identification by adopting domestic laws addressing this issue. The EU thus leaves to EU member states the final decision as to render re-identification unlawful.[51] However, this does not imply that the CJEU offers no guidance into the potential unlawfulness of re-identification schemes.

### 3.5 Evaluating the Unlawfulness of Re-identification Means

In *Breyer*, the Court underlined – as it has been previously stated - that lawfulness is a criterion to be taken into account when assessing the reasonable means likely to be used to re-identify data. The Advocate General even noted that "[i]t is irrelevant, in that context, that access to the personal data is possible *de facto* by infringing data protection laws.[52]" What is interesting here is that the Advocate General seems to consider that being able in practice to re-identify data may be a violation of data protection law: within the context of Directive 95/46 (the EU Data Protection Directive), the practical possibility of accessing personal data must be considered reasonable only if it is done

---

[47] Data Protection Act 2018, Explanatory Notes, Commentary on provisions of the act, § 492.

[48] The Guardian, https://www.theguardian.com/society/2009/oct/16/pro-life-alliance-abortion-jepson-case, last accessed 2024/04/29.

[49] Department of Health, R (on the application of) v. Information Commissioner, England and Wales High Court (Administrative Court), 20 April 2011.

[50] See recital 149 of the GDPR.

[51] It is important to acknowledge that certain practices can be deemed unlawful without necessarily falling under the scope of criminal law. This acknowledgment is particularly relevant in the context of the GDPR, which primarily emphasises administrative and civil measures to ensure data protection. Under the GDPR, the focus extends beyond criminal prohibitions to encompass a broader legal framework. For example, a breach of data protection principles, such as inadequately anonymising personal data, can result in significant administrative fines and sanctions imposed by data protection authorities.

[52] CJEU, OPINION OF ADVOCATE GENERAL CAMPOS SÁNCHEZ-BORDONA delivered on 12 May 2016, Case C-582/14 Patrick Breyer v. Bundesrepublik Deutschland.

through lawful means. In other words, any means of access to personal data must comply with applicable data protection laws and regulations.

The Advocate General emphasises that the requirement for access to be reasonable inherently implies that it must be lawful. This means that even if there are practical methods to access personal data, such access would not be considered reasonable if it involves infringing data protection laws. It is thus irrelevant whether access to personal data is possible in practice through methods that violate data protection laws. Even if such unauthorised access methods exist, they cannot be considered a reasonable means of access under Directive 95/46.

Second, the Court emphasises that the legal means are manifested by the existence of legal channels, which supposes that there must be legal provisions allowing a specific person to get the information needed to re-identify data.

Eventually, the Court is interested in the means used to re-identify, and not by the re-identification by itself. This solution has been reiterated by the General Court[53] in the SRB vs EDPS case.[54]

From this perspective there are still uncertainties with regard to this "legal means" criterion: does it mean that there are unlawful means? If so, what are they? Does it mean that, positively, there must be legal channels, explicit provisions that enable us to collect additional information? In other words, is re-identification expressly allowed or at least possible?

One might even wonder whether, in practice, the "lawfulness" of re-identification is really a relevant criterion to increase the level of protection of personal data. From this perspective, the Swedish Data Protection Authority (DPA) has cautioned against interpreting the concept of personal data in a manner that excessively limits the scope of protection, as it would significantly weaken the overall protection offered by the GDPR. The Swedish DPAs view can be read as follows:

"[a]n interpretation of the concept of personal data that means that it must always be demonstrated that there is a legal possibility to link such data to a natural person would, according to IMY, entail a significant limitation of the regulation's scope of protection, and open up opportunities to circumvent the protection in the regulation. This interpretation would, among other things, be contrary to the purpose of the regulation as set out in Article 1(2) of the GDPR.[55]"

Specifically, the Swedish DPA's interpretation suggests that data can be considered anonymised—and thus not subject to GDPR restrictions—even if there is a risk that the data can be re-identified using sophisticated techniques. This broad interpretation allows for greater access to such "anonymised" data, which could then be used in ways that might not fully protect individuals' privacy.

---

[53] CJEU, JUDGMENT OF THE GENERAL COURT (Eighth Chamber, Extended Composition), in Case T-557/20, Single Resolution Board (SRB) v. European Data Protection Supervisor (EDPS), 26 April 2023.

[54] Lodie A., Are personal data always personal? Case T-557/20 SRB v. EDPS or when the qualification of data depends on who holds them., European Law Blog, (2023).

[55] IMY, Supervisory decision under the General Data Protection Regulation Tele2 Sverige AB's transfer of personal data to third countries, DI-2020-11373, 30 June 2023.

The main concern here is about the robustness of the anonymisation standards being applied. If data that can be re-identified is still treated as anonymised, the protections that the GDPR aims to provide might be undermined. This means that individuals' personal information could be exposed or misused, despite the intention to keep it private. The core issue is that re-identification techniques are advancing, and what might be considered anonymised today could become identifiable tomorrow. Therefore, the interpretation by the Swedish DPA raises questions about whether current anonymisation practices are sufficient to safeguard personal data in the long term.

This would impose a strict requirement that may exclude certain types of data from being considered personal data, even if they pose potential risks to individuals' privacy. Entities may exploit loopholes by structuring their data practices in a way that avoids meeting the strict legal criteria for personal data, as they would all outside the scope of the GDPR.

Consequently, the purpose of the following section will be precisely devoted to analysing the compliance of re-identification schemes with the GDPR. More specifically, as underlined previously, we will try to figure out whether re-identification for research purposes can be deemed lawful under EU data protection law just like UK lawmakers expressly enshrined.

## 4  Re-identification Under the GDPR in Practice

In this section we will try to provide some insights on the way re-identification schemes can be deemed compliant with the GDPR. Although it does not constitute a handbook for practitioners or researchers willing to carry out re-identification, we underline some contentious points that should be taken into account.

### 4.1  Re-identification as Data Processing

The first question that one has to address is whether re-identification is subject to the GDPR, or more generally, to EU data protection law. The GDPR applies materially to "the processing of personal data.[56]" Re-identification, to be subject to the GDPR, must constitute data processing.

Under the GDPR, data processing involves (among other actions) collecting, consulting or using data.[57] From this perspective, carrying out re-identification research should be considered as data processing since researchers will at least consult and store the data once re-identified. The purpose of this data processing operation lies in the scientific progress that computer scientists accomplish by discovering new weaknesses of a de-identification technique used to release a dataset publicly or to protect data.

From this background, the Norwegian DPA claims for instance that "(i)f someone should succeed in re-identifying the data, and this results in personal data being processed, the organisation responsible for the data must assume the role of data controller

---

[56] See Article 2 of the GDPR.

[57] See Article 4 (2) of the GDPR.

for them.[58]" Re-identification, and the subsequent storing, sharing or re-use of data must be considered data processing for which the organisation re-identifying data assumes the role of data controller.

This conclusion seems to be in line with Article 4 (7) of the GDPR[59] since researchers re-identifying data determine the purposes and means of data processing. Indeed, they use re-identification tools to reveal anonymisation vulnerabilities which may lead to massive data breaches. The organisation the researcher works for could be considered as a data controller in this context as well, as the French DPA (CNIL[60]) suggested.[61]

For instance, the University employing a computer scientist to work on re-identification issues can be considered as a data controller, but we will not discuss this issue further.

When a researcher re-identifies pseudonymised data, it is obvious that data are being processed since said researcher receives personal data (pseudonymised data) and further processes them in order to re-identify them.

## 4.2 Contentious Points Relating to Re-identification for Research Purposes with Regard to Data Processing Principles

Since computer scientists re-identifying data should logically be considered as data controllers,[62] they must comply with at least one of the legal bases provided for in the GDPR. Indeed, the first principle relating to data processing as laid down in Article 5 of the GDPR is that "data shall be (…) processed lawfully.[63]" The existence of a legal basis is not enough to process data lawfully pursuant to the GDPR but it remains a salient issue since it reveals all the difficulties that may arise when it comes to re-identification.

As a preamble, it is worth mentioning that in this situation, the purpose of the data processing operation is a scientific purpose since the main aim is to reveal data security vulnerabilities and thus protect data subjects' personal data and privacy. However, the scientific purpose is not, by itself, a legal basis. In other words, processing data for research purposes, or any other "legitimate" purpose does not mean that such processing is lawful under the GDPR or benefits from a legal basis.

From this background, the French DPA released guidelines on the legal regime applicable to data processed for scientific purposes. The guidelines identify as possible legal bases, the consent of subjects, the performance of a task carried out in the public

---

[58] Datatilsynet, https://www.datatilsynet.no/en/regulations-and-tools/reports-on-specific-sub jects/anonymisation/, last accessed 2023/01/25.

[59] See Article 4 (7) of the GDPR.

[60] The French Data Protection Authority (CNIL) (Commission Nationale de l'Informatique et des Libertés) has been a critical player in the landscape of data protection, particularly with the enforcement of GDPR in France.

[61] CNIL, https://www.cnil.fr/sites/cnil/files/atoms/files/consultation_publique_-_presentation_ du_regime_juridique_applicable_aux_traitements_a_des_fins_de_recherche.pdf, last accessed 2023/01/25.

[62] Cf above, Subsect. 4.1.

[63] See article 5 of the GDPR.

interest, or the legitimate interest pursued by the controller.[64] However, we will see that each of these candidates may involve interpretation issues or do not fit the reality of re-identification attacks carried out for scientific purposes.

First, consent is not likely to be a relevant legal basis for processing data when launching a re-identification attack. As a matter of fact, researchers carrying out such an attack are unaware of who the data subjects were initially, since the main aim of their operation is to try to identify data subjects from de-identified datasets. Such a legal basis is thus inoperative to provide a clear framework.

CNIL's guidelines also mention the performance of a task carried out in the public interest. However, once again, it is unclear whether such a legal basis is fit for purpose in such a scenario. More specifically, the GDPR requires that "(w)here processing is carried out in accordance with a legal obligation to which the controller is subject or where processing is necessary for the performance of a task carried out in the public interest or in the exercise of official authority, the processing should have a basis in Union or Member State law.[65]" In other words, member states' law should expressly contain provisions regarding data processing carried out by researchers willing to re-identify data. The question goes as to whether this law must be specific or whether a general statute of researchers under domestic law could be sufficient as well as a mere transposition of the GDPR into domestic law.[66]

For instance, the French "Loi informatique et libertés" transposing the GDPR into French law contains some provisions which can be interesting when considering the legal basis to process data in the public interest. Article 78 of the law provides that "A decree [...] shall determine under what conditions and subject to what safeguards the rights provided for in Articles 15, 16, 18 and 21 of the same Regulation may be waived in whole or in part with regard to processing for scientific or historical research purposes, or for statistical purposes.[67]" Said decree has been adopted and it interestingly provides that when processing data for scientific purposes, data controllers and processors shall "respect the rules of ethics applicable to their sectors of activity.[68]" Although such texts do not expressly state that computer scientists carrying out re-identification attacks for scientific purposes can rely upon the public interest legal basis, they are worth mentioning since they deal with the obligations of data controllers in the field of research.

It is undeniable that computer scientists performing a re-identification attack contribute to the progress of research in the field of cybersecurity and data protection. As such, researchers, as members of a public-funded institution, placed under the authority

---

[64] CNIL, https://www.cnil.fr/sites/cnil/files/atoms/files/consultation_publique_presentation_du_regime_juridique_applicable_aux_traitements_a_des_fins_de_recherche.pdf, last accessed 2023/01/25.

[65] See recital 45 of the GDPR.

[66] See for instance French Loi n° 78-17 du 6 janvier 1978 relative à l'informatique, aux fichiers et aux libertés, art 78 and Décret n° 2018-687 du 1er août 2018 pris pour l'application de la loi n° 78-17 du 6 janvier 1978 relative à l'informatique, aux fichiers et aux libertés, modifiée par la loi n° 2018-493 du 20 juin 2018 relative à la protection des données personnelles, art. 100-1.

[67] Loi n° 78-17 du 6 janvier 1978 relative à l'informatique, aux fichiers et aux libertés.

[68] Décret n° 2019-536 du 29 mai 2019 pris pour l'application de la loi n° 78-17 du 6 janvier 1978 relative à l'informatique, aux fichiers et aux libertés.

of the Ministry of Higher Education and Research, can be considered as exercising a task carried out in the public interest.

The last legal basis likely to authorise researchers to undertake a re-identification attack is the legitimate interest of the data controller. However, this legal basis seems to be inoperative for our case study since "this basis applies only to private entities.[69]" Indeed, recital 47 of the GDPR provides that "(g)iven that it is for the legislator to provide by law for the legal basis for public authorities to process personal data, that legal basis should not apply to the processing by public authorities in the performance of their tasks.[70]"

Actually, things are much more nuanced since it depends on the legal status of universities in EU Member States. Can a university be considered a public authority? As already mentioned, this is the case in France, but the status of researchers and research bodies may vary from one EU Member State to another. Besides, even when researchers are employed by a public authority, legitimate interest would be excluded as a valid legal basis only when processing is carried out in the performance of their tasks. One may consider that, since research is the core function of researchers and research public bodies such as universities, the legitimate interest would not be a valid legal basis in the context of computer scientists undertaking re-identification attacks for research purposes.

The French DPA has clarified the use of the "legitimate interest" legal basis by public bodies to process data. It underlined that "[t]he GDPR provides that the legal basis of legitimate interest does not apply to processing carried out by public authorities in the performance of their tasks". "However, […] this provision does not prevent the use of this legal basis when the processing, although necessary for its current administration or operation, does not fall within the strict performance of its tasks as provided for by the texts.[71]" This tends to exclude legitimate interest from the scope of the legal bases likely to be used for research in the field of re-identification.

Interestingly, some universities have published their own guidelines to specify under what grounds their agents could process data. For instance, the University College London (UCL) claims on its official website that "(a)s a public authority, most of UCL's processing will be undertaken using Article 6(1)(e) above, the 'public task' condition. This applies when the processing is necessary for UCL to perform a task in the public interest. Examples include most of UCL's research, teaching and learning activities – we can clearly demonstrate a 'public task' basis for these because performing such tasks is a core part of UCL's Charter and Statutes".[72]

In the same line of thought, the French "CNRS" (National Centre for Scientific Research) published guidelines on GDPR compliance in the field of research. These guidelines are quite similar to the UCL ones since the CNRS claims that "[i]n the context of research activities, processing should preferably be carried out on the basis

[69] Maldoff G., How GDPR changes the rules for research, (2016).

[70] See Recital 47 of the GDPR.

[71] CNIL, https://www.cnil.fr/fr/les-bases-legales/choisir-base-legale, last accessed 2023/04/13.

[72] University College London, Practical Data Protection Guidance Notice: Legitimate interests as a lawful basis for processing personal data, https://www.ucl.ac.uk/data-protection/gui dance-staff-students-and-researchers/practical-data-protection-guidance-notices/legitimate, last accessed 2023/01/25.

of consent (respecting the principle of informational self-determination), but processing may also be based on the exercise of a public interest mission.[73]"

The 'public interest' legal basis seems, therefore, to be relevant when considering the re-identification of anonymised datasets carried out by computer scientists in the fulfilment of their tasks.

However, the lawfulness of processing is not the only core principle which is likely to raise issue when it comes to re-identification.

### 4.3  Re-identification with Regard to GDPR's Core Principles: Transparency and Data Minimisation

One may also question the compliance of re-identification attacks as regards the principle of transparency as provided for in Article 5 of the GDPR.[74] Re-identification is by its very nature a covert data processing operation since, when processing data, researchers do not know who the data belong to, so they cannot be transparent on the way they process data vis-à-vis data subjects.

In addition to the transparency principle, re-identification challenges the principle of data minimisation as well. This principle mandates that data controllers limit the collection, processing, and retention of personal data to what is strictly necessary for the intended purpose. However, this principle may conflict with the objectives of re-identification attacks conducted for scientific research, which aim to uncover vulnerabilities in anonymisation techniques by analysing extensive datasets. As a matter of fact, the very idea of re-identification involves cross-referencing data by collecting vast amounts of data to single out data subjects, to link some attributes to them or infer information.[75]

### 4.4  Re-identification and Data Subjects' Rights

One of the main issues with regard to the compliance of re-identification attacks with the GDPR is the exercise of data subjects' rights. Data subjects have a right to be informed of the processing of their data, a right to object to such processing or even in some situations a right to erasure.

In particular, in the scenario that we are considering, it seems very difficult for data controllers (researchers) to comply with the right of data subjects to information as provided for by Article 14 of the GDPR.[76] Indeed, when re-identifying an anonymised

---

[73] InSHS IAP and others, Les Sciences Humaines et Sociales et La Protection des Données à Caractère Personnel Dans Le Contexte de La Science Ouverte: Guide Pour La Recherche. (2023).

[74] See Article 5 of the GDPR.

[75] Yang, H., Yi, D., Liao, S., Lei, Z., & Li, S., Cross Dataset Person Re-identification. In ACCV Workshop., (2015).

[76] Article 14 of the GDPR reads as follows: '1. Where personal data have not been obtained from the data subject, the controller shall provide the data subject with the following information:(a) the identity and the contact details of the controller and, where applicable, of the controller's representative;(b) the contact details of the data protection officer, where applicable;(c) the

dataset, researchers do not know who the data subjects are, so they cannot inform them about the data processing carried out.[77] They can only inform them *a posteriori*, which is not the right way to proceed since "(n)otice should be provided at the time when the data is first collected, and it must include the controller's identity and contact information.[78]"

However, Article 14 paragraph 5 provides for some exemptions, in particular, data controllers do not have to provide a notice when the situation makes it impossible or too complex. The same goes when such a requirement is likely "to render the processing impossible or seriously impair the achievement of the objectives of that processing.[79]"

By virtue of this article, computer scientists undertaking re-identification attacks would not be constrained to inform people since it would be impossible as they do not know the exact nature of the data processed, nor who the data subjects actually are.

Furthermore, under the GDPR, data subjects benefit from other rights concerning the processing of their data.[80] However, when data processing for scientific purposes is involved, data controllers may be exempted from compliance with these obligations.[81] It means that researchers undertaking a re-identification attack would not have to protect all these rights. Nonetheless, these exemptions must be provided by European or domestic law, besides they are not absolute and "must be necessary for the fulfilment of [the research] purposes.[82]" Indeed, these derogations must be interpreted narrowly and the research project must comply with the GDPR in other respects.[83]

These rights include the right to access their personal data[84] held by data controllers, allowing them to verify the lawfulness of the processing. Additionally, data subjects have the right to rectify inaccurate or incomplete personal data,[85] ensuring the information held about them is accurate and up-to-date. Furthermore, individuals can request the erasure of their personal data under certain circumstances, commonly referred to as the

---

purposes of the processing for which the personal data are intended as well as the legal basis for the processing;(d) the categories of personal data concerned;(e) the recipients or categories of recipients of the personal data, if any;(f) where applicable, that the controller intends to transfer personal data to a recipient in a third country or international organisation and the existence or absence of an adequacy decision by the Commission, or in the case of transfers referred to in Article 46 or 47, or the second subparagraph of Article 49(1), reference to the appropriate or suitable safeguards and the means to obtain a copy of them or where they have been made available.'

[77] For instance, computer researchers cannot reasonably inform data subjects in a research context when they analyse historical medical records studying effects of certain treatments from last century. While data subjects may still be alive today, locating and contacting them individually is practically impossible due to the incomplete nature of the records, changes in contact information etc.

[78] Maldoff G., How GDPR changes the rules for research, (2016).

[79] See Article 14 § 5 (b) of the GDPR.

[80] See in particular Article 15, 16, 17, 18, 20, 21 of the GDPR.

[81] See Article 89 of the GDPR.

[82] Maldoff G., How GDPR changes the rules for research, (2016).

[83] Office of the Data Protection Ombudsman, Rights of the data subject in scientific research, https://tietosuoja.fi/en/rights-of-the-data-subject-in-scientific-research.

[84] See Article 15 of the GDPR.

[85] See Article 16 of the GDPR.

"right to be forgotten.[86]" They also have the right to restrict processing,[87] to object to processing, and the right to not be subject to a decision based solely on automated processing.[88]

Eventually, the way re-identification operates can contradict the philosophy of the GDPR which is to ensure privacy by design and by default.

### 4.5   Re-identification and Privacy by Design and Default

Similarly, privacy by design and default, another key aspect of the GDPR highlighted in Article 25, mandates that data protection measures be integrated into the design and operation of systems, ensuring that privacy is maintained by default. While data protection by design involves integrating privacy concerns during the whole lifecycle of a product or service,[89] privacy by default "refers to the selection of the most privacy friendly configuration by default.[90]". It is clear from what has been stated above that re-identification runs contrary to the very nature of the principles of privacy by design and privacy by default.

A re-identification scheme is by default aimed at cross-referencing and collecting as much information as possible to succeed in re-identifying data subjects. Such a system is invasive and intrusive by design and default.

These issues underscore the complexity surrounding the intersection of data protection principles and re-identification activities for scientific purposes. While re-identification may serve legitimate research objectives and public interest missions, reconciling these activities with fundamental data protection principles remains a legal grey area. Clarity is needed regarding the extent to which derogations from data subjects' rights are permissible for researchers conducting re-identification attacks and whether specific legal provisions or broader research statutes suffice to authorise such activities. Addressing these concerns is essential for ensuring compliance with the GDPR while facilitating valuable scientific research.

## 5   Conclusion

Re-identification has considerably developed over the years, posing significant risks for individuals' privacy and data protection. The proliferation of data-driven technologies and the widespread collection of personal information for public re-use and marketing purposes have intensified these privacy risks. While re-identification techniques offer insights into the effectiveness of data de-identification methods and risk mitigation strategies, they also raise complex legal questions, be it for data controllers, computer scientists, researchers and users. In particular, some of the requirements laid down by

---

[86] See Article 17 of the GDPR.

[87] See Article 18 of the GDPR.

[88] See Article 22 of the GDPR.

[89] Jasmontaite L., Kamara I., Zanfir-Fortuna G., Leucci S., Data Protection by Design and by Default: Framing Guiding Principles into Legal Obligations in the GDPR. European Data Protection Law Review, 4, 2, 168–189 (2018).

[90] Ibidem.

the GDPR seem to be hard to meet for researchers willing to carry out research in the field of re-identification.

To address these challenges, we propose several guidelines and future directions. Firstly, European Data Protection Authorities and institutions should release specific guidelines on the lawfulness of re-identification for research purposes, including defining the scope of permissible research activities and the conditions under which re-identification is lawful. Legally, clear definitions of research scope can help ensure that data usage is limited to what is necessary, minimising the risk of re-identification. Secondly, implementing robust technical measures to protect data, including encryption, anonymisation, pseudonymisation and access controls is crucial. Thirdly, developing and enforcing organisational policies could improve data protection and could ensure compliance with data minimisation and privacy by design and by default principles.

**Acknowledgments.** This work has been supported by the ANR 22-PECY-0002 IPOP (Interdisciplinary Project on Privacy) project of the Cybersecurity PEPR and by Inria action-exploratoire DATA4US. **The authors would like to warmly thank Cedric Lauradoux, researcher at Inria and general co-chair of the APF conference, for his insights on the technical part of the work and for his help.**

**Disclosure of Interests** The authors are part of the Privatics team, Inria and work in close collaboration with Cédric Lauradoux, co-chair of the APF conference, on a regular basis.

# References

1. Cohen, A.: Attacks on Deidentification's Defenses. In: 31st, USENIX Security Symposium (USENIX Security 2022), pp. 1469–1486 (2022)
2. De Montjoye, Y.-A., Hidalgo, C.A., Verleysen, M., Blondel, V.D.: Unique in the crowd: the privacy bounds of human mobility. Sci. Rep. **3**, 1376 (2013)
3. Dhondt, K., et al.: A run a day won't keep the hacker away: inference attacks on end-point privacy zones in fitness tracking social networks. In: CCS 2022: Proceedings of the 2022 ACM SIGSAC Conference on Computer and Communications Security, pp. 801–814 (2021)
4. Esayas, S.Y.: The role of anonymisation and pseudonymisation under the EU data privacy rules: beyond the 'all or nothing' approach. Eur. J. Law Technol. **6**(2), 19 (2015)
5. Finck, M., Pallas, F.: They who must not be identified—distinguishing personal from non-personal data under the GDPR. Int. Data Priv. Law **10**(1), 35 (2020)
6. Gambs, S., et al.: De-anonymization attack on geolocated data. J. Comput. Syst. Sci. **80**, 1597 (2014)
7. Garfinkel, S.: De-Identification of Personal Information, NIST Interagency/Internal Report (NISTIR), National Institute of Standards and Technology, Gaithersburg, MD (2015)
8. Giomi, M., Boenisch, F., Wehmeyer, C., Tasnádi, B.: A unified framework for quantifying privacy risk in synthetic data. Proc. Priv. Enhanc. Technol. **2023**(2), 312–328 (2023). https://doi.org/10.56553/popets-2023-0055
9. Henriksen-Bulmer, J., Jeary, S.: Re-identification attacks—a systematic literature re-view. Int. J. Inf. Manage. **36**, 1184–1192 (2016)
10. Jasmontaite, L., Kamara, I., Zanfir-Fortuna, G., Leucci, S.: Data protection by design and by default: framing guiding principles into legal obligations in the GDPR. Eur. Data Prot. Law Rev. **4**(2), 168–189 (2018)

11. Kikuchi H., et al.: Ice and fire: quantifying the risk of re-identification and utility in data anonymization. In: 2016 IEEE 30th International Conference on Advanced Information Networking and Applications (AINA), Crans-Montana, Switzerland, pp. 1035–1042 (2016)
12. Lodie A.: Are personal data always personal? Case T-557/20 SRB v. EDPS or when the qualification of data depends on who holds them. European Law Blog (2023)
13. Lodie, A.: Case C-479/22 P, Case C-604/22 and the limitation of the relative approach of the definition of 'personal data' by the ECJ. European Law Blog (2024)
14. Maldoff, G.: How GDPR changes the rules for research (2016)
15. Mitchum, R.: New Kind of Attack Called "Downcoding" Demonstrates Flaws in Anonymizing Data (2022)
16. Mourby, M., et al.: Are 'pseudonymised' data always personal data? Implications of the GDPR for administrative data research in the UK. Comput. Law Secur. Rev. **34**, 222–233 (2018)
17. Narayanan, A., Shmatikov, V.: How To Break Anonymity of the Netflix Prize Dataset (2006)
18. Peloquin, D., DiMaio, M., Bierer, B., et al.: Disruptive and avoidable: GDPR challenges to secondary research uses of data. Eur. J. Hum. Genet. **28**, 697–705 (2020)
19. Rocher, L., et al.: Estimating the success of re-identifications in incomplete datasets using generative models. Nature Commun. **10**, 3069 (2019)
20. Rogers, J., Song, A.: Digital marketing in the legal profession: what's going on and does it Matter? Law Technol. Hum. **5**(2), 134–164 (2023)
21. Schermer, B.W.: The limits of privacy in automated profiling and data mining. Comput. Law Secur. Rev. **27**, 45–52 (2011)
22. Sheth, J., Charles, H.: Next Frontiers of research in data driven marketing: will techniques keep up with data tsunami? J. Bus. Res. **125**, 780–784 (2021)
23. Shokri, R.: Membership Inference Attacks Against Machine Learning Models (2017)
24. Spindler, G., Schmechel, P.: Personal data and encryption in the European general data protection regulation. JIPITEC **7**, 164 (2016)
25. Stalla-Bourdillon, S., Knight, A.: Anonymous data V. personal data—a false debate: an EU perspective on anonymization, pseudonymization and personal data. Int. Data Priv. Law **10**(1), 38 (2020)
26. Sweeney, L.: K-anonymity: a model for protecting privacy. Int. J. Unc. Fuzz. Knowl.-Based Syst. **10**(5), 557–570 (2002)
27. Sweeney, L.: Simple demographics often identify people uniquely. Carnegie Mellon University, Data Privacy Working Paper 3. Pittsburgh (2000)
28. Willemson, J.: Fifty shades of personal data – partial re-identification and GDPR. In: Gryszczyńska, A., Polański, P., Gruschka, N., Rannenberg, K., Adamczyk, M. (eds.) APF 2022. LNCS, vol. 13279, pp. 88–96. Springer, Cham (2022). https://doi.org/10.1007/978-3-031-07315-1_6
29. Hu, Y., Yi, D., Liao, S., Lei, Z., Li, S.Z.: Cross dataset person re-identification. In: Jawahar, C.V., Shan, S. (eds.) ACCV 2014. LNCS, vol. 9010, pp. 650–664. Springer, Cham (2015). https://doi.org/10.1007/978-3-319-16634-6_47
30. Zuiderveen Borgesius, F.: The Breyer case of the court of justice of the European Union: IP addresses and the personal data definition. Eur. Data Prot. Law Rev. **3**(1), 130–137 (2017)
31. TheNewYorkTimes:        https://www.nytimes.com/2006/08/09/technology/09aol.html. Accessed 09 June 2023

# How to Drill into Silos:
# Creating a Free-to-Use Dataset of
# Data Subject Access Packages

Nicola Leschke[1]([⊠])[iD], Daniela Pöhn[2][iD], and Frank Pallas[1][iD]

[1] Paris Lodron Universität Salzburg, Salzburg, Austria
{nicola.leschke,frank.pallas}@plus.ac.at
[2] Universität der Bundeswehr München, München, Germany
daniela.poehn@unibw.de

**Abstract.** The European Union's General Data Protection Regulation (GDPR) strengthened several rights for individuals (data subjects). One of these is the data subjects' right to access their personal data being collected by services (data controllers), complemented with a new right to data portability. Based on these, data controllers are obliged to provide respective data and allow data subjects to use them at their own discretion.

However, the subjects' possibilities for actually using and harnessing said data are severely limited so far. Among other reasons, this can be attributed to a lack of research dedicated to the actual use of controller-provided subject access request packages (SARPs). To open up and facilitate such research, we outline a general, high-level method for generating, pre-processing, publishing, and finally using SARPs of different providers. Furthermore, we establish a realistic dataset comprising two users' SARPs from five services. This dataset is publicly provided and shall, in the future, serve as a starting and reference point for researching and comparing novel approaches for the practically viable use of SARPs.

**Keywords:** Data subject access request · DSAR · GDPR · Personal data access · Privacy · Transparency · Data Portability · Personal data package

## 1 Introduction

The General Data Protection Regulation (GDPR) [18] was implemented to strengthen and harmonize data protection within the European Union (EU). Besides other principles [36], it particularly clarifies and details, as per Art. 15, the data subjects', i.e., individuals', right to request information about how personal data concerning them are being processed and used. In addition, data subjects have the right to request a copy of their data from data controllers by performing data subject access requests (DSARs). The information has to be provided in an electronic and understandable format within a month and free

© The Author(s), under exclusive license to Springer Nature Switzerland AG 2024
M. Jensen et al. (Eds.): APF 2024, LNCS 14831, pp. 132–155, 2024.
https://doi.org/10.1007/978-3-031-68024-3_7

of cost, with few exceptions. Similar to this electronic copy, data subjects have the right to receive a machine-readable copy of the data *they provided* according to Article 20 GDPR (data portability). Even though both Articles include the right to get a copy of personal data and are therefore subsumed under the term "ex-post transparency" [34], the content and format of those copies differ from a legal point of view. Nevertheless, the interfaces found in practice do not necessarily make such a difference [15]. Therefore, we define the electronic data received via a respective ex-post transparency request as *subject access request package (SARP)*, which can be a copy of personal data according to Art. 15, a data portability archive according to Art. 20, or a general SARP according to alike regulations.

As one of the most comprehensive legislative changes introduced with the GDPR, the strengthened right to data access combined with the newly introduced right to data portability have been broadly studied in matters of their effects on users and services [31,51,54]. Compared to related obligations regarding transparency [20,22,23], however, explicitly technical contributions concerning Art. 15 and 20 GDPR are rather rare.

There might be different reasons for this, particularly including a lack of publicly available datasets of SARPs as the basis for conducting respective research. In practice, such datasets can only be generated through access requests from actual accounts filled with realistic data. The datasets do, in turn, inevitably include personal data and using them for research purposes is only possible after properly de-identifying them. Still, a risk of re-identification remains [43], especially when SARPs corresponding to the same data subject but originating from different controllers are published. Therefore, researchers either publish only selected information [10] or decide to keep the SARPs private [41,48,51].

To overcome these issues, the overall goal of this work is to provide a method to generate publicly available SARP datasets spanning both multiple data subjects and multiple controllers, thereby facilitating further research towards technical contributions to ex-post transparency. Beyond that, SARP datasets might also fuel research in other, non-privacy related areas, like social network analysis [60]. Hence, we discuss the method to generate SARPs based on research-only accounts and de-identification, create a corresponding dataset, and compare our approach with related work. Our contribution is therefore two-fold:

1. We propose a method to receive SARPs and curate them into a dataset specifically tailored to answer research questions within different areas and
2. We provide an initial, minimal dataset spanning two data subjects and comprising particularly relevant services, namely Apple, Amazon, Facebook, Google, and LinkedIn.

These contributions unfold as follows: First, in Sect. 2, we present the goals of our work. Next, in Sect. 3, we outline and discuss our method to create and fill the accounts, as well as to de-personalize the SARPs before the analysis. In Sect. 4, we subsequently describe the actual creation of the datasets. We discuss related work in Sect. 6 before we conclude the paper.

## 2    Goals

DSARs and their resulting SARPs should allow data subjects to get a deeper understanding of the processing and, thereby, foster informed privacy decisions, which are a core concept in modern privacy legislation. In practice, however, the underlying rights are underused and not perceived as useful [49,54]. To combat this lack of usefulness, there is an inherent need for research in the area of DSARs, ranging from conceptual research to the design and development of user-centric tools and applications [21,41,49]. Moreover, researchers from other domains like social network analysis [11,56,60], or even (mental) health research [26,53] are increasingly interested in using SARPs.

One current hindrance for respective research is the lack of publicly available, realistic, and controlled reference datasets spanning multiple controllers, which could serve as a playground for development and as comparative baseline. Instead, most researchers resort to individual ad-hoc data acquisition [3,41,49]. Generating such datasets is challenging due to, e. g., ethical and privacy-related considerations [35,52,60], or the (un-) willingness of real data subjects to share their personal data for research purposes [47].

Having a public dataset that allows for the initial exploration of SARPs, in turn, could fuel research, especially in the privacy domain, but also beyond (see, e. g., studies on the subject of risk-based authentication [33]). Once existing, such a dataset can be used for different research directions, ranging from subject-centered research (e. g., focused on analyzing behavior over time) and policy-focused questions (exploring, for instance, the impact of regulatory changes or data interoperability as a prerequisite for a useful right to data portability) to social science [11], mental health [53], or geoinformatics [28]. Similarly, research focused on gaining insights into the operation and behavior of controllers (see Cambridge Analytica [30]) or on reverse engineering the inherent schema of a controller (e. g., to facilitate completeness checks or as a prerequisite for data interoperability) would also benefit from a properly shaped realistic and publicly available SARP dataset.

Against this background, we strive for the following goals to be achieved by the data generation method and the respective initial dataset proposed herein:

- **G1: A public, free-to-use dataset of SARPs.** In order to facilitate future research in the area of DSARs, we identify the need for a publicly available dataset containing exemplary SARPs. The dataset shall be free-to-use for anyone and any purpose. Therefore, real personal data is unsuitable, as it would be subject to the GDPR and might only be processed for specific purposes.

  A so-shaped dataset will, on the one hand, lower the entry barriers for researchers by providing an initial overview and allowing for the exploration of different research questions referring to different aspects of DSARs. On the other hand, the envisioned dataset shall also serve as a reference to compare different SARP-related approaches and applications (e. g., personal information management systems, privacy dashboards, or data portability concepts like the Data Transfer Project[1]).

– **G2: Machine-readable data.** Automation is at the core of the envisioned processing of the SARPs and the respective research to be fostered. This calls for machine-readable (preferably structured or semi-structured) data.[2]

– **G3: Detailed data.** The SARP data to be provided shall be as detailed and complete as possible to unlock a broad variety of analyses and facilitate the application of a broad variety of research approaches. This shall benefit privacy-related analyses (e. g., to assess the completeness of the data) but also detailed examinations beyond the privacy domain.

– **G4: Realistic data.** The data themselves should be as realistic as possible. In particular, they should resemble all characteristics, peculiarities, and possible intricacies to be expected for real-world SARPs as provided by relevant controllers. Otherwise, they would miss crucial aspects, potentially leading to unusable tools and misleading research.

– **G5: Controlled data.** To allow for the sound assessment of the results from automated analyses, to lower the risk of re-identification after publication, and to limit the amount of multi-party data[3], the data should be dedicatedly generated in a controlled fashion, with as little uncontrolled factors as possible.

– **G6: Two-dimensional analysis of the dataset.** Specifically, we strive for a dataset that enables both **(6a) cross-controller** and **(6b) cross-subject** analysis. Unlocking the real value from the right of access (RtA) for data subjects calls for the possibility of integrating SARP data from different controllers and analyzing them conjointly (6a). At the same time, researchers (e. g., from social sciences) are interested in analyzing patterns within a specific controller but across subjects (6b). Both types of analysis and respective research shall thus be supported by the dataset.[4]

– **G7: A method to facilitate the repeated creation of such datasets.** Simply providing *one fixed* dataset will hardly suffice for a wide variety of possibly intended SARP data analyses. Long-term studies (e.g., for measuring the effect of new regulations), as well as foreseeable changes on the side of data controllers, might, for instance, call for repeated data collection. In contrast, topic-specific studies may necessitate the inclusion of particular silos (sets of) controllers. To support such additional dataset creations and relieve respective researchers and developers from repeatedly having to identify crucial

---

[1] https://github.com/google/data-transfer-project, last accessed 04/10/2024.

[2] Potential options for data extraction from unstructured SARP data – e. g., based on advanced AI/ML approaches – are considered out of scope herein and shall thus not be considered further.

[3] Data concerning multiple data subjects, which might not be willing to share their data for research purposes.

[4] Last but not least, this will also contribute to the development of tools and respective research in the field of data portability through, e.g., the reverse engineering of controller-specific data schemas and respective matching concepts.

challenges and respective countermeasures, we also intend to explicate our dataset creation method for flexible reuse.

## 3   General Method and Precursory Considerations

Recently, and closest to our work, Boeschoten et al. [10] introduced a workflow and corresponding software to enable the automated collection of SARPs "donated" by real users. In order to protect the privacy of those users, Boeschoten et al. [9] cleaned the data before usage as the participants shared their full name, phone number, and email address. While this approach is useful in, for example, social science research to meet **G3** and **G4**, it does not properly address the remaining goals pursued herein.[5] We, therefore, provide a similar method, albeit with the difference of not using real data of natural persons, but rather employing pseudonymous accounts specifically tailored for research.

In particular, our approach rests on the following precursory considerations:

- **C1: Research-only Accounts.** Within such research-only accounts, there are different levels of artificial creation and usage patterns: real data, pseudonymized, or entirely artificial. Accounts with real data might give the most realistic SARPs. Having several accounts with actual data at one provider might have implications, such as data aggregation, inference with personal contacts, or a restriction of the number of accounts in terms of services (ToS). Pseudonymized accounts reduce the linkage to the actual data. Further identifiers could be included within the derived and observed data, so a complete pseudonymization might not be possible. Based on our experience, artificial intelligence (AI) pictures are promising, though services might try to combat these types of pictures. Profile generators provided inconsistent data when the profile data was non-US-based. If the account is deactivated, reactivating it might be more challenging. Even if the profile does not impersonate a natural person, fake accounts might be against the ToS. Therefore, depending on the controllers and the aim of the study, dedicated research-only accounts with real data of the researchers might be the only option. In those cases, the data has to be monitored closely and extensive data cleansing has to be done before publishing. In general, we recommend using pseudonymous accounts. To avoid detection, respect the terms of the controllers, and satisfy **G4**, we stay close to the truth.
- **C2: Data Provided.** The datasets received by GDPR Article 15 or Article 20 DSARs should contain all the data subjects *provided* beforehand. In addition, requests according to Article 15 shall also contain *derived and observed* data but do not have to (but sometimes nonetheless might) be provided in a machine-readable form (see **G2**). Automated post-processing or analysis,

---

[5] In particular, the data is not controlled by the researchers (**G5**). Additionally, the re-identification risks increase when adding multiple sources, so **G6a** can better be fulfilled with non-personal accounts.

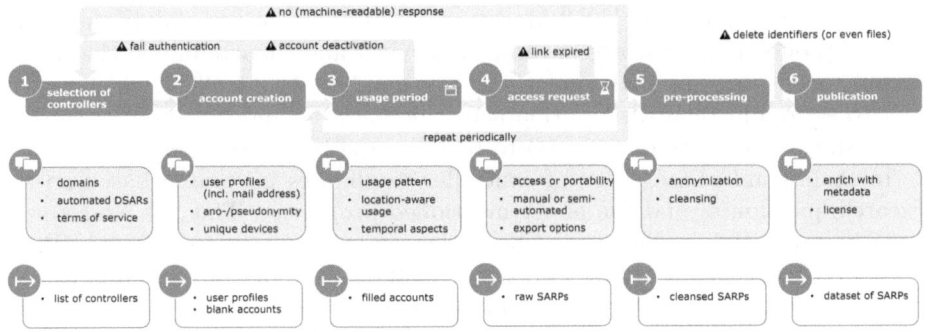

**Fig. 1.** Method for creating a dataset of SARPs.

as ultimately envisioned herein, can, in consequence, only be guaranteed for subject-provided data obtained via Article 20.

This subject-provided data can be controlled (**G5**) during account creation and usage in terms of amount and type of data shared, and their completeness in the SARP can be assessed. However, to allow for meaningful (legal) analysis, the data derived and observed holds high value. Therefore, we tend to use Art. 15 requests, but choose machine-readable options where available.

– **C3: Datasets of Different Sizes.** Depending on the use case, the appropriate dataset size differs. For making generalizations, large datasets are required. Also, for machine learning applications, as used, for example, in [28], big datasets are mandatory to avoid overfitting [42]. In contrast, small datasets are well-suited for human exploration.

The method proposed hereafter should thus allow for the creation of small, human-understandable datasets but should also facilitate large-scale data generation. This can be achieved in three different dimensions: adding more controllers, adding more users, or repeating the access request periodically.

Building on these precursory considerations, we outline our method to generate a dataset containing a sample of SARPs from research-only accounts. The method – visualized in Fig. 1 – contains six steps, each with specific necessary decisions, potential failures, and resulting outputs, as delineated below.

### 3.1 Selection of Controllers

In general, the selection of controllers that should be included in the dataset depends on the research question but also has some implications on a meta-level: First, there are two important properties of the controllers to be examined, independently from the research question: a controller has to be subject to (one or many) DSAR obligations because of processing personal data and being subject to regulations like the GDPR, CCPA, or the like, and the DSAR processing should be automated. The latter is crucial in order to not raise unnecessary burdens to the controllers [49].[6] Moreover, it is important to validate the legitimacy of pseudonymous usage in terms of services and (depending on the study) possibly overriding national regulations. The number of controllers, which we will refer to as $n$, and their distribution across domains should be chosen according to the specific research question(s) to be answered by the dataset.

In the second phase, a list of controller candidates is curated based on the research question(s) and the above-mentioned properties. This list is narrowed down by an initial exploration of both the request process and the SARPs to be expected for getting a first impression of the suitability of the dataset. This can be done by analyzing related datasets or using the personal accounts of the researchers. The request process should be evaluated based on scientific reports or other sources (privacy policy, third-party instructions, exploration using personal account), assuming that there is a web-based request process and that the controller-sided process is automated as well [41]. The resulting candidate list should be discussed in terms of suitability for answering the research question. Then, the fixed number of $n$ controllers is chosen from the candidate list.

After finishing the step, there should be a list of $n$ suitable controllers to be included in the study. Possible failures in this phase occur when there is not enough information for the initial exploration of the request process and expected outcome. Additionally, this step might also be a good point to get feedback from an ethics commission.

## 3.2 Creation of Accounts

Given the list of controllers, the next step is the creation of user accounts. As outlined before, and to meet **G5**, our method is here specifically tailored for pseudonymous research-only accounts. Therefore, within this step, the first action item is the creation of user profiles. This user profile has to include identification data, specifically a name, and, as the creation of an online account mostly relies on email addresses, an email alias, which should both be pseudonymous. In order to pass automated verification and depending on the domain, there might be the need to provide a credit card. This will have to be real data. Even more, some controllers require a unique phone number or even uniqueness of devices. Hence, there might be a need for a dedicated device and/or SIM card. Some controllers might also ask for physical addresses, which have to be within the territorial scope of meaningful access rights, e. g., the EU or California. Other profile data, such as date of birth, sex, education history, working history, or domain-specific information, can be entirely artificial. Overall, this makes the user profile as a whole pseudonymous and not entirely artificial.

As laid down above, one email address per user profile is required for identification purposes. Hence, the first account to be created is an email account. Using this email account, one account per controller and user profile is initiated.

---

[6] The likelihood of a well-defined and automated access process is higher for big companies.

As a result, $m$ user profiles and $n * m$ blank accounts are created. If the user profiles fail the account verification of a specific controller, one should step back and choose another controller in Sect. 3.1.

## 3.3   Usage Period

Within a well-defined usage period, the blank accounts are filled with data by using the controllers' services as normally and realistically as possible. Like creating accounts, data can be generated with real, pseudonymized, or artificial data. As discussed before, the variants have different implications. One exception is multi-party data, such as in private messages and chats. Here, the privacy of other people has to be taken into consideration. Options include chatting with other participating accounts or alternatively faking chats using, for example, AI text-generation tools.

Users may use services differently, ranging from the device (smartphone or notebook, different operating system, and browser) used for accessing to usage patterns (for example, before/after work and privacy-concerned) and frequency (low frequency vs. power user). Different types of usage may be better suited depending on the research questions. The practicability for the researchers is another issue. For example, a smartphone may require a SIM card or can only be used with WiFi. In order to prevent too many ties to a natural person, usage is recommended only in public networks and not when traveling in private life. This usage pattern may result in the risk of forgetting to use the accounts, so implementing reminders is an important part of the study. This is especially important when the research question requires a long usage period, for example when recurring DSARs are to be performed to allow for the analysis of SARP development.

At the end of the usage period, the accounts are filled with usage data. However, during usage, an artificial account might be detected and deactivated. In that case, real verification data is beneficial. If the account can not be reactivated, return to step 2.

## 3.4   Data Subject Access Requests

To receive a SARP containing the usage data, the next step deals with DSARs. As mentioned above, the controllers within the study should allow for web-based DSARs. Still, the process of executing DSARs is often tedious and error-prone [13]. Leschke et al. [32] automate the user side of the request process, an approach that can prove beneficial for more extensive studies. After sending the request, researchers should regularly check the progress of the DSAR. Although the compilation of SARPs can take up to 30 days, it is often much faster, especially with automated processes. This unpredictable time frame is particularly challenging, as some controllers provide the SARPs only for a few days. If that timespan is passed, a new DSAR has to be triggered [41]. In order to meet **G3**, the DSAR should be based on Art. 15 GDPR ("data access"), as those might hold substantially more data than a portability request according to Art. 20.

However, some web forms do not allow one to choose between different legal bases. Often, users are allowed to choose between several request options, e. g., a timeframe, categories of data, or even the format of the SARP.

As a result, the researchers have gathered $n*m$ SARPs that contain account and usage data. However, there are still some controllers that are not responding to access requests or provide no machine-readable data, some even send a paper copy via mail. In this case, a DSAR based on Art. 20 should be used. Only in worst-case scenarios, one has to go back to step 1.

### 3.5    Pre-processing and Cleansing

As hinted with **C1**, the data exports may contain personal identifiers and other possibly identifying data, such as IP addresses, geolocations, phone numbers, and names. Boeschoten et al. [11] analyze the possibly identifying data (namely email, name, phone, URL, username, and faces) of Facebook, WhatsApp, Snapchat, and Twitter before presenting an approach for de-identification. However, the data and data structure may change, and the approach is limited to the websites presented. The amount of possibly identifying data depends on the preceding considerations and the level of pseudonymization. Pseudonymous data is vulnerable to re-identification attacks. Hence, the data has to be pre-processed and cleansed before it can be analyzed. By analyzing the data for possibly re-identifying information and using different methods to anonymize them, re-identification is made reasonably unlikely and the dataset can be assumed to be non-personal data [19]. A list of typical re-identifying information can be found in Table 1. When other anonymization techniques fail, e. g., when having personal data within unstructured files like PDF and media files, the corresponding personal data or even file has to be deleted. This should be noted down and a list of deleted items should be provided as supplementary material.

However, further data can also call for proper cleansing. Concerning privacy, this particularly includes multi-party data – such as messages and emails – birthdays, addresses, internet service providers (ISPs), and user IDs. Further data included in the SARP may put the conducting researchers at risk in matters of security. For instance, information about device features may be used by adversaries to mimic the researchers' identity and try to attack them on that basis (see [16]). Our list of to-be-cleansed data laid out in Table 1 thus also includes data items without direct privacy-related relevance. As a result, a set of cleansed SARPs that can be used for analysis and publication is provided.

### 3.6    Publication

It is important to provide supplementary information with the dataset to make it meaningful to the public. Particularly, the goals, methods, and pre-processing steps – with a focus on anonymization techniques used and the list of deleted items – are important additional information (see **G5**).

**Table 1.** Data with privacy- and security-related relevance.

| Type | Privacy | Security |
|------|---------|----------|
| Device | ID | ID |
|  | Name | Name |
|  |  | Type |
|  |  | Hardware |
|  |  | Serial number |
|  |  | OS version |
| Network | Type | Type |
|  | ISP | ISP |
|  | IP address | IP address |
|  | Location | Location |
| Website | User ID | User ID |
|  |  | Sessions and cookies |
|  |  | Authentication methods |
|  |  | User agent |
| User | Name | Account name |
|  | Email address | Email address |
|  | Address | Address |
|  | Birthday | Birthday |
|  | Phone number | Phone number |
|  | Media | Media |
| Multiparty | Usernames, User IDs | Usernames, User IDs |
|  | Timestamps |  |
|  | Message content | Message content |

With the pseudonymous accounts, location-aware usage and dedicated pre-processing steps that are inherent within this method, the risks of re-identification are minimized. However, as there are different attack vectors on datasets [59], and the risk of re-identification within social media data is high [14], the risk cannot be ruled out. The original users (researchers) must be aware of the risks and benefits of publication and should, therefore, explicitly consent to its publication for all research purposes.

The concrete publication modalities should be discussed. The options include providing the dataset for public download or making it available upon request. While the former removes all barriers to research, the latter lowers the risks of misuse.

## 4   Initial Application: A Minimal SARP Dataset

Following our method, we create a dataset to meet our goals defined in Sect. 2. We decide on a micro-dataset to allow for manual exploration. Hence, the number of controllers and user profiles should be minimal but sufficient to reach goal **G6**. Addressing **G6a**, we set the number of controllers to five. A meaningful cross-controller analysis should cover both, controllers within a specific domain (that intuitively should have similar data) and controllers from different domains (to explore domain-independent data). To meet **G6b**, we include two user profiles. In the following, we present the specific choices that were made to create such a dataset using the method presented in Sect. 3.

## 4.1 Selection of Controllers

We concentrate on the major online services of Google, Apple, Facebook, and Amazon (GAFA), as well as LinkedIn as a business social network. All of these are available for citizens within the EU and, thereby, subject to the GDPR. As major services, they are all viewed as gatekeepers according to the Digital Markets Act (DMA) and, therefore, of particular interest to researchers and the general public.

Previous work [41,49,50] indicates that those controllers meet **G2** and offer web-based access requests [32], which lowers the burden on the controller (see Sect. 3.1). With Facebook and LinkedIn, there are two services from the social network domain. Google serves as the e-mail provider, but is also an active player in several domains, of which we choose video streaming (via YouTube). Within the same domain, we use Apple and Amazon as video streaming services. To demonstrate their inherent complexity, the former is additionally viewed as an app store provider, while the latter is used for online shopping.

## 4.2 Creation of Accounts

We created two pseudonymous user profiles, which we refer to as user $A$ and $B$, and respective research-only accounts for the selected websites. While both profiles can overall be viewed as pseudonymous, especially verification data had to have some connection to real data, which we monitored closely in order to de-identify them during data cleansing (as described in Sect. 3.4).

Using the information of the user profiles, two Gmail accounts where opened first. Next, each Gmail address was used to generate an Apple ID. To preserve the uniqueness of devices and use Apple's app store, we used the iOS apps of the covered services. Each user profile was bound to a dedicated iPhone that was solely utilized for this study. After installation, the Facebook, Amazon, and LinkedIn apps were used to create accounts based on the Gmail address and the corresponding user profile.

Wherever possible, we stuck to the truth, mimicking realistic data. For example, Apple required billing information for one of the created accounts. There, we used the university's address. For the other account, another email address (university email address) and phone number were good enough to pass account verification. In order to get realistic data (**G4**), as provided by real data subjects, we allowed cookies and other tracking methods where they did not contradict the pseudonymous usage, e. g., we allowed location-based services only while using the apps.

Table 2. Characterization of the users.

| Characteristics | Users |
| --- | --- |
| Type of profile | Research only |
| Usage pattern | Infrequent usage |
| Websites | Amazon, Apple, Facebook, Google, LinkedIn |
| Privacy attitude | Allow tracking |
| Actions | Search, like, save, hear/watch, chat |
| Device | iPhone |

### 4.3 Usage Period

In order to keep the accounts both realistic (**G4**) and controlled (**G5**), we agreed on the information we wanted to share, the kind of interactions to be performed, and the close and frequent communication. During the usage period, we took notes about the privacy-relevant decisions that were made.

For social networks, we agreed on the following actions. We followed accounts of huge organizations, but allowed only interaction with non-personal accounts, meaning either legal bodies or other research-only accounts. Public posts are liked and might be commented on. Chats are allowed between research-only accounts. Images might be uploaded if they convey no personal data, e. g., show objects without humans or are artificially generated. For video streaming, we watched free videos with ads and interacted with them (i. e., pausing, fast-forwarding or -backwarding, and stopping). For online shopping, we mainly used the search function and added items to the wish list. Similarly, the app store search function was used, and free apps were installed.

Overall, we paid special caution regarding multiparty data, particularly messages and emails. These should not be made public unless all parties know that the content will be used for this study. Thus, we interacted only with the other research-only accounts. During pre-processing, all incoming messages had to be reviewed in detail, and, when in question, be deleted. For all other data categories, we tried to fill the accounts with as much data as possible to get realistic data.

With the purpose of pseudonymous usage, we limited the usage to mainly the universities' WiFis and other public and open networks. Hence, the usage was restricted to office hours, leading to infrequent users. Nonetheless, with creating a dataset for (privacy) research in mind, this limitation was acceptable. The usage period for users $A$ and $B$ was two and four months, respectively. The characteristics of the accounts are summarized in Table 2.

### 4.4 Data Subject Access Requests

After a few weeks of usage, we requested the corresponding subject access request packages via web interfaces. When given the choice, we chose to request the data according to Art. 15 GDPR. Where possible, we chose a machine-readable format (e. g., selecting JSON instead of HTML) to reach **G2**. In general, we decided to include as much data as possible, setting the export options to all data (time and category-wise) to address **G3**.

For successful requests, the SARPs were available within six days. In one case (Apple), the initial request failed for one account completely.[7] After 30 days, a second request was made, choosing only selected data. For the second account, the response was incomplete. LinkedIn provided one basic export very quickly (within 20 min), and a 'complete' export within two days. In general, a few issues appeared during the process, such as unusual waiting time, no notification email though the SARP was downloadable, and expired download possibilities.

### 4.5   Pre-processing and Cleansing

As stated in Sect. 3.5, possibly identifying data within the SARPs needs to be obfuscated or removed in order to de-identify the dataset. While the labels of such identifying information vary across controllers, we hereafter refer to the general concepts that need special caution. An extensive list of all identifying data, their paths, and the methods that are used to de-identify them is provided with the dataset. Per user, we de-identified the data in the same manner across providers, e.g., replacing the original mail address with *firstname.lastname@gmail.com*, to preserve implicit cross-provider linkages. Across users, however, we employed different de-identifications to reflect the variety of available approaches.

First, we looked at the content of machine-readable files, which are of particular interest for future work. For string-based identifiers like name, username, and mail, we chose two approaches: user $A$s data was obfuscated by using semantic descriptions (like *Firstname Lastname*), while the corresponding data of user $B$ was replaced by the common substitute *\*\*\**. Usage dates are preserved, but the birthdate was set to a random date between 1960 and 2000. As the addresses provided were the addresses of an institution, this data did not demand obfuscation. However, observed location data, like the usage area, are generalized (e.g., from district to city). To preserve the variety of IP addresses found in the SARPs, within the data of user $A$, we replaced every unique IP address with a random IPv4 loopback address.[8] To make the data more realistic, we replaced all the IP addresses of user $B$ with the static IP 77.24.117.117 that occurred in the dataset. For multiparty data, such as chats within the Facebook and LinkedIn SARPs and emails in Google's SARPs, we preserve only the messages between the research-only accounts. Other data is deleted to protect the sanctity of mail.

Device-specific security-related information is also obfuscated. Here, we do not want to preserve the variety and, hence, performed the same steps for both users: All numbers within device identifiers, like version and operation system,

---

[7] We did not receive a response within 30 days.

[8] This allows the address to be in range 127.0.0.0–127.255.255.255.

are replaced with *0*. The internet service provider (ISP) listed in the dataset was replaced by *unknown*.

Next, we looked at unstructured files. Media files, like pictures, could contain personal data. However, as the accounts were filled with the publication in mind, we never provided personal pictures and were also cautious with the metadata of the media files provided. We found personal information within PDF files provided by Amazon. Despite the fact that PDFs can be edited, the result may be reversed. Therefore, we decided to provide only screenshots of the edited content (see Fig. 2).

Finally, we examined the SARPs as a whole. Some folders and filenames included identifying information, e. g., the Facebook SARPs contained the username as the directory name. Here, we chose the same approach and replacement strategies as within machine-readable data. All de-identification steps are recorded and provided with the dataset as supplementary material.

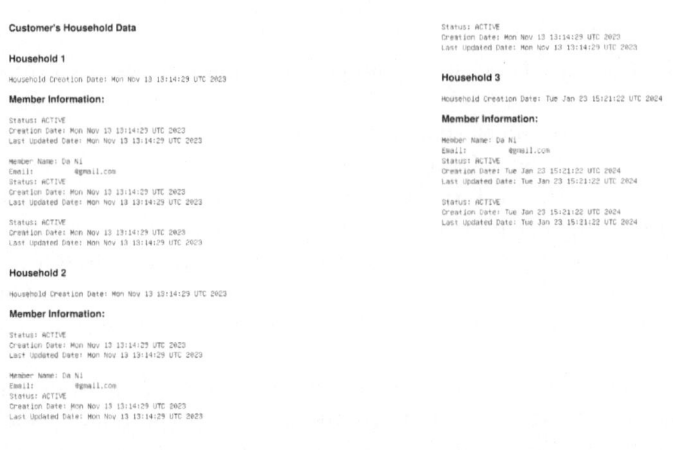

(a) First page of the PDF.          (b) Second page of the PDF.

**Fig. 2.** Screenshot of two pages with information about the data subject's household provided by Amazon.

## 4.6    Publication

In order to allow meaningful analysis or development of user-centric applications, we enrich the dataset with metadata and provide a description of the method and its implementation within the dataset generation as laid down above. Additionally, an enumeration of files changed and screenshots of de-identified PDFs before deletion are provided. We make the dataset, including the cleansed SARPs and additional information, available for other researchers.[9] The dataset is provided under a Creative Commons Attribution 4.0 license.

# 5 Dataset Description and Preliminary Analysis

In this section, we describe the dataset generated in Sect. 4 and perform a preliminary analysis for select examples to demonstrate its usefulness.

## 5.1 Characterization of the Dataset

A summary of the characterization of the SARPs for both users can be found in Table 3. While usage time and export options denote important metadata to contextualize the files, the other columns describe the cleansed SARPs. As stated in Sect. 4.4, LinkedIn provided two separate exports, which we also describe separately. Apple and LinkedIn provide solely machine-readable data (particularly tabular data in CSV format). For Apple, this can be explained by a lack of unstructured data provided to the controller (no profile picture or similar), but on the other hand, it means that no metadata or explanations are provided. LinkedIn, in turn, provides download links for media files within one dedicated CSV file. Authenticated users must thus follow this link in a subsequent step to access this data. For Amazon and Google, the ratio of machine-readable (semi-structured) files is around 70% and for Facebook even less. This is due to media or domain-specific files (like calendar data as ICS) and contextualizing files like readmes. Those contextualizing files are provided in the language chosen during usage (German in our case), while labels within semi-structured files are either in English or German (or abbreviations). Google also adapts folder and file names to the chosen language.

In general, the number of files may vary between subjects depending on the time and type of usage. We observe that the file structure (number of files) is adapted accordingly. Interestingly, the file type chosen within the export options also makes a difference (see Facebook). The general file structure, precisely the varying depth across subjects, indicates that semantics/context is given not only within files but also in the folder structure.

## 5.2 Preliminary Analysis

To demonstrate its usefulness, we perform an exemplary manual analysis. Along this line, future research can use the provided dataset for exploration and as a baseline for evaluating user-centric applications for such SARPs. We select the topic of ads that were also targeted by Pöhn and Gruschka [40] for Twitter (now X). Possible data comprises the targeting audience, shown ads, clicked ads,

---

[9] Access to the dataset can be requested through https://doi.org/10.5281/zenodo.11634938.

**Table 3.** Characterization of the SARPs for subject $A$ and $B$.

| Prov. | Usage time (in months) | Export options | File types | # Subfolders | # Files | Export size |
|---|---|---|---|---|---|---|
| Amazon | 2/4 | all categories | CSV (32/49) EML[a] (2/5) JPEG (1/2) JSON (3/3) PDF (9/10) TXT (4/4) | 41/49 | 51/73 | 1.2 MB/1.4 MB |
| Apple | 2/4 | all data , max. 1 GB/max. 4 GB | CSV (8/3) | 20/1 | 8/3 | 71.8 KB/294.8 KB |
| Facebook | 2/4 | all data, JSON, on my computer | JSON (39/0) HTML (0/63) TXT (29/28) JPG (0/4) PNG (1/15) GIF (7/7) | 45/76 | 76/117 | 12.3 MB/13.5 MB |
| Google | 2/4 | all data, frequency once, ZIP, 4 GB | HTML (8/11) CSV (10/13), JSON (27/28) TXT (14/14) PDF (1/1) MBOX (1/1) VCF (1/0) ICS (1/0) README (1/1) JPG (0/2) | 44/51 | 64/71 | 1.54[b] MB/1.2 MB |
| LinkedIn | 2/4 | all data | CSV (18/21) | 0/0 (part 1/2) 0/0 (part 1/2) | 13/18 19/21 | 3.9 KB/6.0 KB 6.2 KB/9.2 KB |

[a] Those files are deleted in the cleansing step, as they contain multiparty data. However, we included them in this list to demonstrate the different formats to be expected within SARPs. Same goes for MBOX and VCF files.
[b] The MBOX file deleted has a size of 7.8 MB, which inflates the raw SARP

and the corresponding metadata (such as location that may hint at location-based ads). First, we analyze a single SARP by focusing on LinkedIn before comparing the analysis with the other online services of that specific user. This step demonstrates the dataset's suitability to reach **G6a**. Finally, to demonstrate **G6b**, we compare the results with the results of the second user.

The SARP of subject $B$ for LinkedIn consists of 21 CSV files, including 'Ad_Targeting', and 'Ads_Clicked'. In 'Ad_Targeting', several organizations are stated in the column of company followers. According to 'Ads_Clicked', subject $B$ has clicked on 98 ads in four months, which seems too high for a user who tries not to click on ads. Unfortunately, only the ID of the ad is included in 'Ads_Clicked'. Thereby, we do not know which ads were marked as clicked. The number of reactions, stated in 'Reactions', such as liking a post (which the user regularly did, following the profile filling strategy), is lower, with 30 in total.

Regarding the SARPs for Google and Apple, no ad-related data was found. This observation is interesting as ads were displayed while watching YouTube videos. In contrast, Amazon shows "Interest based ads" according to the file 'Advertising.OptOut.csv'. According to Amazon, nine ads were streamed within Amazon Music. However, no ads related to Prime Video could be found, although ads were shown according to subject $B$. Also, no data about clicked ads or the targeting group is included. This contrasts with Facebook, where the two interacted ads are shown, and the target group consists of sales and mead (sic!). However, Facebook does not include data about ads that they have shown.

When comparing the ad-related data of subject $B$ with those of subject $A$, we also noticed several ads for the latter, again with the amount of clicked (28) exceeding the one of reactions (11). Like for $B$, the file 'Ad_Targeting' seems misaligned. For example, company industries have the entry of '2025'. In contrast to subject $B$, the member age and selected further data are not given. The most significant contrast between $A$ and $B$ can be found with Amazon's SARPs. While Amazon does not provide more detail on ads for subject $B$, it does so for subject $A$ with advertiser audiences (70 in total), advertiser clicks (3), and Amazon audiences (9 categories).

These (due to space constraints) rudimentary findings should already illustrate the kind of research and insights possible with the kind of datasets proposed and provided herein. Further, more detailed insights would go beyond the scope of the paper at hand and are planned for a subsequent publication.

## 6    Related Work

Following the introduction of the GDPR, the CCPA, and similar legislation, DSARs gained significant research attention throughout recent years. Large portions of respective research so far regard real-world DSAR processes, ranging from their empirical examination and analysis [12,17,40,41,49–51] to the identification of generalized user journeys [39]. Common findings include a limited understandability of provided data [13,54] and difficulty in actually executing DSARs in the first place [37,39]. Proposals for addressing these include data dashboards [6,29,44,50] and personal data information management systems (PIMS) [4,57,58]. With a more pragmatic focus on the actual givens, easing the DSAR execution through established means of web automation has also been proposed [32]. None of these endeavors, however, focuses on the content of the provided SARPs or the creation and provision of respective datasets.

SARPs themselves, in turn, have been analyzed from a single controller's point of view, e. g., Twitter (now X) [40] or Instagram [38], and from a comparative perspective [31,41,51]. SARP-focused user studies, in turn, repeatedly conclude that the provided raw data are hardly useful for data subjects [12,54]. Instead, SARP data needs to be automatically pre-processed and analyzed – also across different providers – in order to be actually useful. So far, however, respective measures for automated pre-processing, -analysis, and exploration are lacking. Insofar, our contributions provided herein complement said work and shall foster the kind of research that helps the above-mentioned research strands to move forward.

As for the use of SARP data in research, Habu and Henderson [25] describe how data subject rights can be used for user-fueled research. For example, social media researchers increasingly rely on SARPs. Boeschoten et al. [7–9,11] propose the usage of donated (and de-identified) SARPs for (social science) research. Similarly, Razi et al. [1,45,46] present a case study on collecting Instagram SARPs for adolescent online risk detection and Zannettou et al. [60] analyze TikTok SARPs to observe the effect of TikTok's recommendations on user engagement. Even though a different line of research than ours, such studies will clearly benefit from our data collection method and the respective clarification of necessary considerations in various ways. At the same time, our work clearly builds upon generalized higher-level insights from such research, particularly with regard to ethical and procedural aspects [27,52].

## 7   Discussion

Based on our method introduced herein, we generated a first minimal dataset suitable for researching and developing novel approaches in the context of data access research and for benchmarking different approaches against each other. Even though tailored to meet the goals outlined in Sect. 2, our method has some limitations.

First, our approach consciously favors controlled data and cross-study value over data realness and volume, contrasting pre-existing approaches based on data donations from real users. More representative and extensive data might be favored depending on the intended use case. However, the pre-processing and cleansing steps then require more work. Still, data donations from real users will provide more voluminous and realistic data, albeit at the cost of controlledness and potential privacy risks laid out above.

Second, depending on the language settings of the device and the accounts, the SARP may contain data in different languages and sizes. Based on the languages and locations, the data may be different. Exploring respective effects and implications is, though, up for future work. In addition, the amount of data may vary between different DSARs. In our case, for instance, $B$ had initially (four months difference) received data from Apple that contained more folders and files, including account information. In addition, the name of the content-holding folder changed between the two respective SARPs. This shows that a long-term study might be interesting.

Lastly, the terms of usage may prohibit users from employing pseudonymized accounts for research. Especially for social media applications, pseudonymous accounts might qualify as fake profiles and may therefore be seen as deceptive usage [24]. Respective rights possibly overruling such terms may vary between countries and decisions [55]. In this regard, the German Psychological Society, the Federation of German Psychologists, and the American Psychological Association declare that no studies based on deception are to be carried out unless deception techniques are justified, for example, by a significant gain in scientific knowledge, and no alternative procedure without deception is available [2,5].

In our case, the avoidance of any interactions with individual non-study users largely eliminates the risk of deceit. Thereby, this procedure can be seen as justifiable as new knowledge is gained.[10] Nonetheless, researchers may risk the deactivation of the research accounts.

## 8    Conclusion

Recent privacy legislation, such as the GDPR or the CCPA, has stimulated research in a broad variety of subjects, particularly privacy statements, data sharing, and consent practices. Compared to these, technical contributions concerning ex-post transparency requests are rarely found. However, the process of the DSARs and the content of the SARPs can provide interesting insights into the behavior and compliance of data controllers and strengthen the data subject's sovereignty. Therefore, we identified the need for publicly available datasets of SARPs that can fuel future research in this area.

One recurring issue with SARP-related research is that these packages include personal identifiers in various forms and at various locations. Hence, the data first needs to be de-identified before analyzing and publishing them in line with best practices for data-driven research. To facilitate such research, we proposed a method to create free-to-use datasets of SARPs based on research goals. We then applied our method to the creation of our exemplary dataset. After characterizing our dataset, we exemplarily analyzed the topic of ads in our dataset. Finally, we contrasted and discussed our approach with related work.

In future work, we plan to extend our dataset by including SARPs from additional providers and countries of origin and by adding more users. We also plan to study SARPs in the long term to recognize changes. The analysis should be extended, for example, by performing analyses for different topics or by exploring different approaches for automation. Further important aspects for future work are applications for SARPs that can be developed using the dataset provided herein. For example, privacy dashboard development could be enhanced by taking our dataset as a use case. More research in personal data interoperability is required to allow integrated dashboards combining SARPs of multiple controllers. Our minimal dataset can be used to explore different data integration approaches on such personal data, which will also benefit research on personal data portability.

Overall, we thus envision our method and minimal SARP dataset to foster research in the domain of data subject access requests in a multitude of ways.

**Acknowledgements.** This publication has been supported by the EXDIGIT (Excellence in Digital Sciences and Interdisciplinary Technologies) project, funded by Land Salzburg under grant number 20204-WISS/263/6-6022.

---

[10] Given the explicit focus on research related to GDPR-based DSAR, resorting to Art. 40 DSA is also not a reasonable option.

# References

1. Ali, S., et al.: Getting meta: a multimodal approach for detecting unsafe conversations within instagram direct messages of youth. Proc. ACM Hum.-Comput. Interact. **7**(CSCW1) (2023). https://doi.org/10.1145/3579608
2. American Psychological Association: Ethical principles of psychologists and code of conduct (2017). https://www.apa.org/ethics/code. Accessed 11 June 2024
3. Ausloos, J., Dewitte, P.: Shattering one-way mirrors - data subject access rights in practice. Int. Data Priv. Law **8**(1), 4–28 (2018). https://doi.org/10.1093/idpl/ipy001
4. Barreau, D.K.: Context as a factor in personal information management systems. J. Am. Soc. Inf. Sci. **46**(5), 327–339 (1995)
5. Berufsverband Deutscher Psychologinnen und Psychologen e.V., Deutsche Gesellschaft für Psychologie e.V.: Berufsethische Richtlinien des Berufsverbandes Deutscher Psychologinnen und Psychologen e.V. und der Deutschen Gesellschaft für Psychologie e.V (2022). https://www.bdp-verband.de/fileadmin/user_upload/BDP/website/dokumente/PDF/Profession/Berufsethik/BER-Foederation-20230426-Web-1.pdf. Accessed 11 June 2024
6. Bier, C., Kühne, K., Beyerer, J.: PrivacyInsight: the next generation privacy dashboard. In: Schiffner, S., Serna, J., Ikonomou, D., Rannenberg, K. (eds.) APF 2016. LNCS, vol. 9857, pp. 135–152. Springer, Cham (2016). https://doi.org/10.1007/978-3-319-44760-5_9
7. Boeschoten, L., Ausloos, J., Möller, J.E., Araujo, T., Oberski, D.L.: A framework for privacy preserving digital trace data collection through data donation. Comput. Commun. Res. **4**(2), 388–423 (2022). https://doi.org/10.5117/CCR2022.2.002.BOES
8. Boeschoten, L., van Driel, I.I., Oberski, D.L., Pouwels, L.J.: Instagram use and the well-being of adolescents: using deep learning to link social scientific self-reports with instagram data download packages. In: Companion Publication of the 2020 International Conference on Multimodal Interaction, ICMI 2020 Companion, p. 523. Association for Computing Machinery, New York (2021). https://doi.org/10.1145/3395035.3425185
9. Boeschoten, L., van den Goorbergh, R., Oberski, D.: A set of generated Instagram Data Download Packages (DDPs) to investigate their structure and content, January 2021. https://doi.org/10.5281/zenodo.4472606
10. Boeschoten, L., et al.: Port: a software tool for digital data donation. J. Open Source Softw. **8**(90), 5596 (2023). https://doi.org/10.21105/joss.05596
11. Boeschoten, L., Voorvaart, R., Van Den Goorbergh, R., Kaandorp, C., De Vos, M.: Automatic de-identification of data download packages. Data Sci. **4**, 101–120 (2021). https://doi.org/10.3233/DS-210035
12. Borem, A., Pan, E., Obielodan, O., Roubinowitz, A., Dovichi, L., Mazurek, M.L., Ur, B.: Data subjects' reactions to exercising their right of access. In: Proceedings of the 33rd USENIX Security Symposium (2024)
13. Bowyer, A., Holt, J., Go Jefferies, J., Wilson, R., Kirk, D., David Smeddinck, J.: Human-GDPR interaction: practical experiences of accessing personal data. In: Proceedings of the 2022 CHI Conference on Human Factors in Computing Systems, CHI 2022, pp. 1–19. Association for Computing Machinery, New York (2022). https://doi.org/10.1145/3491102.3501947
14. Branson, J., Good, N., Chen, J.W., Monge, W., Probst, C., El Emam, K.: Evaluating the re-identification risk of a clinical study report anonymized under ema policy 0070 and health Canada regulations. Trials **21** (2020)

15. Bufalieri, L., Morgia, M.L., Mei, A., Stefa, J.: GDPR: when the right to access personal data becomes a threat. In: 2020 IEEE International Conference on Web Services (ICWS), pp. 75–83 (2020). https://doi.org/10.1109/ICWS49710.2020.00017
16. Campobasso, M., Allodi, L.: Impersonation-as-a-service: characterizing the emerging criminal infrastructure for user impersonation at scale. In: Proceedings of the 2020 ACM SIGSAC Conference on Computer and Communications Security, CCS 2020, pp. 1665–1680. Association for Computing Machinery, New York (2020). https://doi.org/10.1145/3372297.3417892
17. Dewitte, P., Ausloos, J.: Chronicling GDPR transparency rights in practice: the good, the bad and the challenges ahead. Int. Data Priv. Law (2024). https://doi.org/10.1093/idpl/ipad026
18. European Parliament and Council: Regulation (EU) 2016/679 of the European Parliament and of the Council of 27 April 2016 on the protection of natural persons with regard to the processing of personal data and on the free movement of such data, and repealing Directive 95/46/EC (General Data Protection Regulation). Off. J. Eur. Union **59**, 1–88 (2016)
19. Finck, M., Pallas, F.: They who must not be identified-distinguishing personal from non-personal data under the GDPR. Int. Data Priv. Law **10**(1), 11–36 (2020). https://doi.org/10.1093/idpl/ipz026
20. Gerl, A., Bennani, N., Kosch, H., Brunie, L.: LPL, towards a GDPR-compliant privacy language: formal definition and usage. In: Hameurlain, A., Wagner, R. (eds.) Transactions on Large-Scale Data- and Knowledge-Centered Systems XXXVII. LNCS, vol. 10940, pp. 41–80. Springer, Heidelberg (2018). https://doi.org/10.1007/978-3-662-57932-9_2
21. Gómez Ortega, A., Bourgeois, J., Kortuem, G.: Personal data comics: A data storytelling approach supporting personal data literacy. In: Proceedings of the XI Latin American Conference on Human Computer Interaction, CLIHC 2023. Association for Computing Machinery, New York (2024). https://doi.org/10.1145/3630970.3630982
22. Grünewald, E., Halkenhäußer, J.M., Leschke, N., Washington, J., Paupini, C., Pallas, F.: Enabling versatile privacy interfaces using machine-readable transparency information. In: Schiffner, S., Ziegler, S., Jensen, M. (eds.) Privacy Symposium 2023, pp. 119–137. Springer, Cham (2023). https://doi.org/10.1007/978-3-031-44939-0_7
23. Grünewald, E., Pallas, F.: TILT: a GDPR-aligned transparency information language and toolkit for practical privacy engineering. In: Proceedings of the 2021 ACM Conference on Fairness, Accountability, and Transparency, FAccT 2021, pp. 636–646. Association for Computing Machinery, New York (2021). https://doi.org/10.1145/3442188.3445925
24. Guo, Z., Cho, J.H., Chen, I.R., Sengupta, S., Hong, M., Mitra, T.: Online social deception and its countermeasures: a survey. IEEE Access **9**, 1770–1806 (2021). https://doi.org/10.1109/ACCESS.2020.3047337
25. Habu, A.A., Henderson, T.: Data subject rights as a research methodology: a systematic literature review. J. Responsible Innov. **16**, 100070 (2023). https://doi.org/10.1016/j.jrt.2023.100070
26. Hafen, E.: Personal data cooperatives – a new data governance framework for data donations and precision Health. In: Krutzinna, J., Floridi, L. (eds.) The Ethics of Medical Data Donation. PSS, vol. 137, pp. 141–149. Springer, Cham (2019). https://doi.org/10.1007/978-3-030-04363-6_9

27. Halavais, A.: Overcoming terms of service: a proposal for ethical distributed research. Inf. Commun. Soc. **22**(11), 1567–1581 (2019). https://doi.org/10.1080/1369118X.2019.1627386
28. Hanny, D., Resch, B.: Clustering-based joint topic-sentiment modeling of social media data: a neural networks approach. Information **15**(4), 200 (2024)
29. Herder, E., van Maaren, O.: Privacy dashboards: the impact of the type of personal data and user control on trust and perceived risk. In: Adjunct Publication of the 28th ACM Conference on User Modeling, Adaptation and Personalization, UMAP 2020 Adjunct, pp. 169–174. Association for Computing Machinery, New York (2020). https://doi.org/10.1145/3386392.3399557
30. Isaak, J., Hanna, M.J.: User data privacy: Facebook, Cambridge Analytica, and privacy protection. Computer **51**(8), 56–59 (2018). https://doi.org/10.1109/MC.2018.3191268
31. Kröger, J.L., Lindemann, J., Herrmann, D.: How do app vendors respond to subject access requests? A longitudinal privacy study on iOS and Android Apps. In: Proceedings of the 15th International Conference on Availability, Reliability and Security, ARES 2020, pp. 1–10. Association for Computing Machinery, New York (2020). https://doi.org/10.1145/3407023.3407057
32. Leschke, N., Kirsten, F., Pallas, F., Grünewald, E.: Streamlining personal data access requests: from obstructive procedures to automated web workflows. In: Garrigós, I., Murillo Rodríguez, J.M., Wimmer, M. (eds.) ICWE 2023. LNCS, vol. 13893, pp. 111–125. Springer, Cham (2023). https://doi.org/10.1007/978-3-031-34444-2_9
33. Makowski, J.P., Pöhn, D.: Evaluation of real-world risk-based authentication at online services revisited: complexity wins. In: Proceedings of the 18th International Conference on Availability, Reliability and Security, ARES 2023. Association for Computing Machinery, New York (2023). https://doi.org/10.1145/3600160.3605024
34. Murmann, P., Fischer-Hübner, S.: Tools for achieving usable ex post transparency: a survey. IEEE Access **5**, 22965–22991 (2017)
35. Ohme, J., Araujo, T.: Digital data donations: a quest for best practices. Patterns **3**(4) (2022). https://doi.org/10.1016/j.patter.2022.100467
36. Pallas, F., et al.: Privacy engineering from principles to practice: a roadmap. IEEE Secur. Priv. **22**(2), 86–92 (2024). https://doi.org/10.1109/MSEC.2024.3363829
37. Petelka, J., Oreglia, E., Finn, M., Srinivasan, J.: Generating practices: investigations into the double embedding of GDPR and data access policies. Proc. ACM Hum.-Comput. Interact. **6**(CSCW2), 1–26 (2022)
38. Peters, Y., Nehls, P., Thimm, C.: Plattformforschung mit Instagram-Daten - Eine Übersicht über analytische Zugänge, digitale Erhebungsverfahren und forschungsethische Perspektiven in Zeiten der APIcalypse. Publizistik **68**(2), 225–239 (2023). https://doi.org/10.1007/s11616-023-00786-8
39. Pins, D., Jakobi, T., Stevens, G., Alizadeh, F., Krüger, J.: Finding, getting and understanding: the user journey for the GDPR's right to access. Behav. Inf. Technol. **41**(10), 2174–2200 (2022). https://doi.org/10.1080/0144929X.2022.2074894
40. Pöhn, D., Gruschka, N.: Past and present: a case study of Twitter's responses to GDPR data requests. In: Rannenberg, K., Drogkaris, P., Lauradoux, C. (eds.) Privacy Technologies and Policy. APF 2023. LNCS, vol. 13888, pp. 57–84. Springer, Cham (2024). https://doi.org/10.1007/978-3-031-61089-9_4

41. Pöhn, D., Mörsdorf, N., Hommel, W.: Needle in the haystack: analyzing the right of access according to GDPR article 15 five years after the implementation. In: Proceedings of the 18th International Conference on Availability, Reliability and Security, ARES 2023. Association for Computing Machinery, New York (2023). https://doi.org/10.1145/3600160.3605064

42. Prusa, J., Khoshgoftaar, T.M., Seliya, N.: The effect of dataset size on training tweet sentiment classifiers. In: 2015 IEEE 14th International Conference on Machine Learning and Applications (ICMLA), pp. 96–102 (2015). https://doi.org/10.1109/ICMLA.2015.22

43. Ramachandran, A., Singh, L., Porter, E., Nagle, F.: Exploring re-identification risks in public domains. In: 2012 Tenth Annual International Conference on Privacy, Security and Trust, pp. 35–42 (2012). https://doi.org/10.1109/PST.2012.6297917

44. Raschke, P., Küpper, A., Drozd, O., Kirrane, S.: Designing a GDPR-compliant and usable privacy dashboard. In: Hansen, M., Kosta, E., Nai-Fovino, I., Fischer-Hübner, S. (eds.) Privacy and Identity 2017. IAICT, vol. 526, pp. 221–236. Springer, Cham (2018). https://doi.org/10.1007/978-3-319-92925-5_14

45. Razi, A., et al.: Sliding into my DMs: detecting uncomfortable or unsafe sexual risk experiences within instagram direct messages grounded in the perspective of youth. Proc. ACM Hum.-Comput. Interact. 7(CSCW1) (2023). https://doi.org/10.1145/3579522

46. Razi, A., et al.: Instagram data donation: a case study on collecting ecologically valid social media data for the purpose of adolescent online risk detection. In: Extended Abstracts of the 2022 CHI Conference on Human Factors in Computing Systems, CHI EA 2022. Association for Computing Machinery, New York (2022). https://doi.org/10.1145/3491101.3503569

47. Skatova, A., Goulding, J.: Psychology of personal data donation. PloS one 14(11) (2019). https://doi.org/10.1371/journal.pone.0224240

48. Sørum, H., Presthus, W.: Dude, where's my data? The GDPR in practice, from a consumer's point of view. Inf. Technol. People 34(3), 912–929 (2021)

49. Syrmoudis, E., et al.: Unlocking personal data from online services: user studies on data export experiences and data transfer scenarios. Hum.-Comput. Interact., 1–25 (2024). https://doi.org/10.1080/07370024.2024.2325347

50. Tolsdorf, J., Fischer, M., Lo Iacono, L.: A case study on the implementation of the right of access in privacy dashboards. In: Gruschka, N., Antunes, L.F.C., Rannenberg, K., Drogkaris, P. (eds.) APF 2021. LNCS, vol. 12703, pp. 23–46. Springer, Cham (2021). https://doi.org/10.1007/978-3-030-76663-4_2

51. Urban, T., Tatang, D., Degeling, M., Holz, T., Pohlmann, N.: A study on subject data access in online advertising after the GDPR. In: Pérez-Solà, C., Navarro-Arribas, G., Biryukov, A., Garcia-Alfaro, J. (eds.) DPM/CBT 2019. LNCS, vol. 11737, pp. 61–79. Springer, Cham (2019). https://doi.org/10.1007/978-3-030-31500-9_5

52. van Driel, I.I., Giachanou, A., Pouwels, J.L., Boeschoten, L., Beyens, I., Valkenburg, P.M.: Promises and pitfalls of social media data donations. Commun. Methods Meas. 16(4), 266–282 (2022). https://doi.org/10.1080/19312458.2022.2109608

53. Verbeij, T., Beyens, I., Trilling, D., Valkenburg, P.M.: Happiness and sadness in adolescents' instagram direct messaging: a neural topic modeling approach. Soc. Media Soc. 10(1) (2024). https://doi.org/10.1177/20563051241229655

54. Veys, S., et al.: Pursuing usable and useful data downloads under GDPR/CCPA access rights via co-design. In: SOUPS @ USENIX Security Symposium (2021)

55. Wauters, E., Lievens, E., Valcke, P.: Towards a better protection of social media users: a legal perspective on the terms of use of social networking sites. Int. J. Law Inf. Technol. **22**(3), 254–294 (2014). https://doi.org/10.1093/ijlit/eau002

56. Wei, M., et al.: What twitter knows: characterizing ad targeting practices, user perceptions, and ad explanations through users' own twitter data. In: 29th USENIX Security Symposium (USENIX Security 2020), pp. 145–162 (2020)

57. Whittaker, S., Massey, C.: Mood and personal information management: how we feel influences how we organize our information. Pers. Ubiquit. Comput. **24**(5), 695–707 (2020)

58. Wilhelm, S., Jakob, D., Gerl, A., Schiegg, S.: Die vision eines personal information management-system (pims) durch automatisierte datenschutzselbstauskunft. In: Daten-Fairness in einer globalisierten Welt, pp. 373–398. Nomos Verlagsgesellschaft mbH & Co. KG (2023)

59. Wong, R.C.W., Fu, A.W.C., Wang, K., Pei, J.: Minimality attack in privacy preserving data publishing. In: Proceedings of the 33rd International Conference on Very Large Data Bases, pp. 543–554 (2007)

60. Zannettou, S., et al.: Analyzing user engagement with tiktok's short format video recommendations using data donations. In: Proceedings of the CHI Conference on Human Factors in Computing Systems (CHI 2424) (2024). https://doi.org/10.1145/3613904.3642433

# Another Data Dilemma in Smart Cities: The GDPR's Joint Controllership Tightrope Within Public-Private Collaborations

Barbara Lazarotto[✉] and Pablo Trigo Kramcsák

Vrije Universiteit Brussel, Brussels, Belgium
barbara.da.rosa.lazarotto@vub.be

**Abstract.** The paper explores the legal challenges and implications of processing personal data within public-private partnerships (PPPs) in smart city projects. Smart cities use a web of technologies to collect and analyse data from various sources, aiming to improve their efficiency and achieve multiple goals. However, this also raises concerns about privacy and adequate protection of data subjects' rights. The paper focuses on the legal basis for data processing regulated by the General Data Protection Regulation (GDPR) and how it differs between the public and private sectors involved in PPPs. The paper argues that the disparity in goals and legal grounds for data processing may create conflicts and uncertainties for joint data controllers, as well as data subjects. It intends to explore potential grey areas in the lawful grounds for collecting data and its implications for citizens' right to data protection in this context.

**Keywords:** Smart cities · Joint controllership · Lawful basis for data processing

## 1 Introduction

Smart cities have represented a contemporary shift in the dynamics of urban living, transforming the way cities function and how their residents engage with their surroundings. These innovative urban landscapes are founded upon an complex web of cutting-edge technologies and methods, that include ubiquitous computing, seamless connectivity, Internet of Things (IoT) devices, big data analytics, and AI-driven decision-making systems. These technologies serve as the backbone, working together to optimise and streamline various facets of a city's operations, ranging from transportation and infrastructure to public safety and city governance. Amidst this technologically advanced ecosystem, data emerges as the quintessential cornerstone for the success and efficacy of smart cities. The tapestry of data woven together from various interconnected sources within the city - sensors, devices, citizen inputs, and administrative databases - forms the bedrock upon which smart cities develop. This wealth of information empowers

M. Jensen et al. (Eds.): APF 2024, LNCS 14831, pp. 156–167, 2024.
https://doi.org/10.1007/978-3-031-68024-3_8

city planners and administrators with insights, enabling them to make proactive, data-driven decisions aimed at enhancing city operation and planning. Hence, smart cities rely on gathering information (both real-time and historical data) from various sources to boost their operational efficiency. This process seamlessly integrates sensory devices into urban structures, creating a sophisticated network of interconnected data collection components. The data-centric model adopted by smart cities has sparked concerns regarding its intrusive nature and the pervasive collection of personal data. Smart cities' systems continuously accumulate a vast range of data points, covering details about citizens' movements, behaviours, preferences, and interactions within the urban environment. Then, this raw data is further processed and analysed to extract useful insights/inferences, make predictions and inform evidence-based decisions. These outputs enable public authorities and service providers to monitor city management and public services, guide operational and strategic initiatives, and achieve other goals. The private sector plays a pivotal role in smart city initiatives, engaging in collaborative endeavours with the public sector through public-private partnerships (PPPs). In these partnerships, private entities often contribute their specialised knowledge in developing/providing ICT infrastructure and software solutions to the public sector. In return, they gain access to the datasets amassed by the smart city network, serving their commercial interests. This symbiotic relationship inherently involves a form of data monetisation (datasets are used to improve current services or generate new revenue streams) integrated into smart city collaborations. The structure of PPPs is not straightforward, demanding a focused understanding of the complexities surrounding personal data processing operations within them, including stakeholders' purposes and various interests involved. The activities carried out by the public sector are fundamentally geared towards serving the public interest and meeting local administrative obligations. This often requires adherence to legal frameworks prioritising the performance of a task carried out in the public interest or the exercise of official authority when handling personal data. In contrast, inherently profit-driven private sector partners tend to base their data processing activities on different legal grounds, such as data subjects' consent or fulfilling contractual obligations. The disparity in goals between the public and private sectors holds significant importance when analysing data processing within PPPs. These partnerships often involve shared control over data, which requires a shared legal grounds for data collection and processing. This requirement prompts crucial inquiries into aligning new business models with applicable legal frameworks. In this context, this study aims to explore the legal challenges and implications of processing personal data within PPPs in smart city projects, exploring the conflicts and uncertainties for joint data controllers when it comes to their purposes, interests and legal grounds for data processing. To do so, this article is organized as follows. Section 2 explores the smart city environment and the private sector's participation in these projects through public-private partnerships and how joint controllership fits into this relationship. The following Sect. 3 investigates the interconnection between public-private partnerships with the concept of joint controllership and the legal grounds for data processing under these circumstances. Section 4 concludes the paper with a critical reflection and the way forward for future research endeavours.

## 2   Smart Cities and the Private Participation

### 2.1   Context

According to studies, by 2050 around 55% of the world's population will live in urban areas, raising multiple challenges to local governments that must handle a series of social, economic, and environmental challenges [1]. It was in this context that the term "smart city" emerged, initially crafted by companies such as IBM and Cisco, later being rapidly adopted by municipalities that aim to use technology as a solution to these challenges while transforming cities into an attractive environment for the development of capital [2].

Although it is beyond the scope of this work to investigate the different concepts of smart cities, it is important to emphasize that there is no consensus on a single definition for this phenomenon. The notion has been through a terminological evolution such as "digital city" [3], "tech city" [4], "wired city [5], "ubiquitous city" [6], and "intelligent city" [7], which is a demonstration of the adoption of technology in cities. For the purposes of this study, we have opted to define "smart cities" as the adoption of cross-sectoral technologies embedded in the urban infrastructure that can be applied in multiple domains such as public transport, traffic management, and urban development solutions both in already existing cities that go through a "smart" transformation or new cities built with "smartness" in mind [8–10].

The first generation of smart cities initially emerged in 2008, posing technology and partnerships with the private sector as a solution to the financial crisis, which restricted considerably governmental budgets [11]. The need for technological solutions opened the market for a new wave of devices and services designed for this purpose, and cities became the center of the new economy, where new solutions for an ecological, sustainable, and socially balanced world could emerge [12].

Personal and non-personal data collection is at the heart of smart cities, collected at an unprecedented granularity by a ubiquitous grid of computing technologies - such as sensors, 5G, blockchain, cloud computing, and IoT amongst others -, spread throughout the urban environment. These data will later be processed and used to feed algorithms that will predict, evaluate, optimize, and manage processes, shaping data-informed decisions and public policy strategies [13,14]. Detailed data would facilitate the formulation of decisions grounded in solid evidence, which ultimately would streamline the provision of public services [2].

In this context, private sector companies play an key role in the development of smart city projects by offering multiple technologies that allow the collection and processing of data, pushing the smart city industry to reach an estimated market size of 3 trillion dollars by 2025 [15]. Multinational corporations such as

IBM, Cisco, and Siemens are examples of this symbiotic relationship between the private and the public sectors which often takes shape through public-private partnerships.

## 2.2  Public-Private Partnerships in Smart Cities

Public-Private Partnerships are a broad category of contracts settled between the public and the private sectors which consist of the sharing of objectives for the delivery of public infrastructure or services. There is no communal European Union definition of PPPs, since they greatly depend of Member States' law framework, yet the European Commission has referred to them as a form of cooperation between public authorities and the private sector with the aim to "ensure the funding, construction, renovation, management or maintenance of an infrastructure or the provision of a service". In this partnership, the private party will provide a service to the public "in the place of" the public partner, though under its control [16].

PPPs became popular in the late 1980s and early 1990s with the influence of New Public Management (NPM) an administrative management doctrine that defended the implementation of market mechanisms to deliver efficiency and quality to the public service while reducing expenses [17,18].

These partnerships can take different forms depending on the legal setting, however, the norm of collaboration between the private and the public sector remains an unchangeable factor. Thus, unlike traditional public contracting, where the government has a privileged position dictating the terms and conditions, PPPs work collaboratively through an agreement between the private and the public parties [19]. Moreover, in PPPs, the private sector may be remunerated on charges levied on the users of the service provided, by regular payments by the public partners, or by a trade-off. Often, in public-private partnerships settled in smart city, the trade-off consists of access and reuse of the data collected for other purposes.

A study commissioned by the European Parliament has found that due to several reasons PPPs are the most preferred contract type for smart city projects [11]. First and foremost, most local governments lack the skills, knowledge, resources, and capacity to address urban challenges and maintain critical infrastructure with restrictive budgets [20]. Additionally, the private sector has extensive expertise offering a wide variety of technologies bringing additional value to these projects, eventually creating the perfect environment for such partnerships [21].

In this context, public-private partnerships have peculiar internal dynamics. Initially, the public sector develops smart city projects having the public interest of developing more efficient urban governance with the help of technology in mind. However, when joining PPPs the private sector has inherently a market-based objective of seeking profit. Thus, there is a visible discrepancy of interests. In traditional PPPs, these discrepancy is often handled, yet when it comes to smart city partnerships it becomes more complex due to technology's embedded capacity that influences how data is generated and further processed. Therefore, by providing technology, the private sector can have a higher capacity to influence the workings of such partnerships embedding its values in smart city projects and shaping the political nature of PPPs [11]. This factor inevitably alters the internal power dynamics of PPPS, giving the private sector the upper hand in internal negotiations due to its technological superiority. These internal power dynamics are relevant since they expand the sphere of influence of the private sector, blurring the divide between the two sectors in reality [22]. They have an impact in the ubiquitous collection of personal and non-personal data by both parties, often in a blurred and complex manner which calls for a joint controllership, a topic which will be further discussed in the next section.

### 2.3    Joint Controllership and Public-Private Partnerships in Smart Cities

The preceding section underscores that in smart city initiatives, public-private collaborations are marked by a heightened complexity. This stems from the overlapping roles of public and private entities in managing both personal and non-personal data, coupled with their divergent objectives. This situation directly raises discussion on the adequate legal basis for processing data in these circumstances and on the controllership of this processing. Since the entry into force of the GDPR, the concepts of controller and processor and their respective roles when processing personal data have been under discussion. The same can be applied to the concept of joint controllership, a topic that has limited scholarly exploration, especially when it comes to its application in the context of public-private partnerships.

It is important to emphasise that not all PPPs will necessarily result in a joint controllership. This assessment must be done in a case-by-case analysis. It is possible that the public sector remains the one that defines the processing activities of the partnership and, therefore, will be considered the data controller, while the private sector will be the processor.

According to Article 26 of the GDPR, joint controllership is defined as where two or more controllers that determine the purposes and means of the processing of personal data. Joint controllership is not determined by the number of entities involved in processing, but by their level of involvement. If two or more actors jointly decide the purpose and the means of processing, they are joint controllers. This determination is flexible, relying on factual circumstances, not on potential contractual agreements among the joint controllers. In this context, according to the European Data Protection Board (EDPB) Guidelines 07/2020 on the concepts of controller and processor in the GDPR, the criterion for joint controllership is the joint participation of two or more entities in the determination of the purposes and means of a processing operation, in a common decision or converging/complementary decisions. It is important to note that according to the Court of Justice of the European Union (CJEU) Judgment in Wirtschaftsakademie, the fact that one of the parties does not have access to personal data that is processed does not exclude joint controllership, given that the purposes and means of the processing might be determined by the two parties. Additionally, the CJEU has clarified that different operators may be involved in different stages of data processing and in different degrees [23,25].

Under the EDPB guidelines, not all partnerships, cooperation, or collaborations result in joint controllership; this necessitates a tailored analysis for each scenario. Occasionally, different entities may independently handle the same personal data within a sequential process, each with their own distinct objectives and methods. Yet, when it comes to public-private partnerships, decisions must be taken by mutual agreement, despite the different interests of the parties - namely the public interest and the economic interest of profit. Therefore, due to the interlinked data processing activities in a smart city context, where the distinction between private and public sector activities frequently becomes indistinct, it could be asserted that scenarios of joint controllership are more probable. This creates a dilemma when establishing a legal basis for data processing since joint controllers must either share the same processing purposes or be closely linked or complementary to each other's commercial interests, according to CJEU case law [26].

## 3   Joint Controllership, PPPs and the GDPR

### 3.1   Previous Considerations

PPPs in smart city settings can potentially involve joint controllership, where multiple actors make common or complementary decisions regarding how data is processed, encompassing both purposes and means. Mere sharing of the same dataset or engaging in collaborative efforts does not automatically imply the existence of joint controllership. However, the concept of joint controllership is expansive, demanding a factual assessment of the comprehensive context in which data processing unfolds, moving beyond formal assessments. This evaluation is crucial to differentiate it from the data controller-data processor relationship delineated in Article 28 GDPR, which may involve processing personal data for another entity's specified purposes and based on its instructions.

Jointly determining the purposes and means of processing does not necessarily mean that controllers share identical purposes for the processing, nor that they exert an equal level of influence over the purposes and means of processing. The key criterion is whether they act in concert or in a complementary manner, rather than independently or unilaterally. Therefore, the lack of access to the dataset by one of the parties or the existence of different grades of control over the shared infrastructure/processing systems are insufficient grounds to dismiss the concept of joint controllership. Joint controllers may process data separately, without any overlap or interaction between their operations. However, they must not have full discretion or autonomy in deciding how the personal data is processed, as this would indicate separate controllership. Joint controllership may involve a complex or multilayered set of data processing activities, each aiming at its own but converging purpose.

Consider a smart parking system designed to optimize parking space usage, alleviate congestion, and generate revenue by collecting and processing data from various sources like sensors, cameras, and mobile applications [23]. This system operates collaboratively among the city, the parking operator, and the technology provider. Each entity maintains distinct purposes for data processing, such as traffic management, service enhancement, or innovation. Despite these individual aims, they harmonize efforts to ensure the system's efficiency and security while aligning their converging objectives. As joint controllers, they collectively establish the purposes and methods of data processing, even if their levels of data access or system control differ.

## 3.2   Joint Controllers' Responsibilities and Compliance

The aim of Article 26 GDPR is to establish transparency and accountability in situations where multiple controllers are involved in processing operations. This is achieved by requiring joint controllers to determine and document their specific responsibilities, as required by Article 26(1), through a formal arrangement, often termed a Joint Controllership Agreement. This agreement outlines the tasks each joint controller is accountable for regarding the different processing activities. Its purpose is to clearly define the allocation of responsibilities for GDPR compliance among the involved parties, with the specific circumstances of each case guiding this determination.

Shared responsibility among joint controllers does not necessarily mean they all have equal involvement in the data processing operations. Nonetheless, how responsibilities are divided within a Joint Controllership scenario, including any agreements about handling liability arising from processing, does not directly impact the rights of the data subjects. Among other effects, data subjects retain the ability to exercise their rights against any of the joint controllers, particularly in cases involving infringements or damages.

To ensure compliance with GDPR provisions regarding joint controllerships, several critical aspects need to be addressed, including the revision of possible contractual relationships with any other parties identified as joint controllers; the identification and assessment of the legal basis for such processing activities; and updating controllers' privacy policies to reflect the joint controllership arrangement.

The responsibilities for complying with GDPR regulations among joint controllers are more extensive than those outlined in Article 26(1). As joint controllers, adherence to GDPR mandates involves implementing data protection principles (Article 5), establishing the legal basis for processing (Article 6), enforcing security measures (Article 32), and notifying supervisory authorities and data subjects about personal data breaches (Articles 33 and 34). Further duties include ensuring compliance with data transfer rules for third countries (Chapter V), and maintaining a record of processing activities, among others.

Additionally, joint controllers are responsible for providing detailed information about joint processing to data subjects. This involves disclosing the purposes and legal basis for processing, the types of personal data involved, recipients or categories of recipients, and the rights of data subjects. These information obligations align with Articles 13 and 14 GDPR.

For data controllers engaged in PPPs, strict compliance with GDPR principles is essential. These principles include, but are not limited to, data minimization, maintaining accuracy, limiting storage duration, ensuring integrity and confidentiality, as well as upholding accountability and transparency. Furthermore, data controllers must uphold data subjects' rights, including access, rectification, erasure, restriction, objection, and data portability, among others. Implementing suitable technical and organisational measures is also important for data security and breach notifications to supervisory authorities and data subjects.

### 3.3 Analyzing and Evaluating Legal Grounds for Data Processing in Joint Controllership Contexts

Joint controllers must carefully determine (and inform to data subjects) the lawful basis for processing joint controlled data. This determination should consider potential interactions between joint controllers and the resulting impact on data subjects. The chosen legal basis must align with the processing's purposes and means while respecting the rights and interests of data subjects. For instance, the public sector might argue that data processing is crucial for tasks in the public interest, like enhancing urban mobility or security. Conversely, the private sector may assert that data processing is essential for contractual obligations, such as providing smart parking or lighting services. Alternatively, the private sector might seek data subject consent, which should be freely given, specific, informed, and unambiguous.

It should be noted that consent might not always serve as a valid or reliable lawful basis for data processing within PPPs due to potential power imbalances or lack of genuine choice for data subjects. Additionally, it might conflict with the public interest or official authority, potentially overriding individual preferences or rights. Therefore, consent should be cautiously used.

Another challenge lies in ensuring that data processing is genuinely necessary for the intended purpose, without reasonable, less intrusive alternatives available. This underscores the need for proportionate data processing that considers impacts on data subjects' rights and interests.

EDPB Guidelines 07/2020 emphasises in Paragraph 164 the need to consider limitations on using personal data for other purposes, depending on the processing and the parties' intentions. In this regard, joint controllers have the duty 'to ensure that they have a legal basis for the processing and that the data are not further processed in a manner that is incompatible with the purposes for which they were originally collected by the controller sharing the data'. 'Each disclosure by a controller requires a lawful basis and assessment of compatibility, regardless of whether the recipient is a separate controller or a joint controller'. Merely having a joint controller relationship doesn't grant the receiving joint controller automatic permission to process the data for additional purposes beyond the agreed-upon scope of joint control.

As per Article 6 GDPR, there are six lawful grounds for processing personal data: consent, contract, legal obligation, vital interests, public interest, and legitimate interests. These grounds are not interchangeable; the choice of lawful basis depends on the specific purpose and context of data processing. Hence, different lawful grounds may apply to various purposes within the same data processing activity, provided these purposes are compatible and consistent. For instance, a bank may process customer data based on a contract to offer banking services and on legal obligation to adhere to anti-money laundering regulations.

However, the GDPR mandates that data controllers inform data subjects about the purposes and lawful grounds for data processing, among other considerations, as this determines the applicable rights and their exercise. Therefore, transparency and accountability are crucial for data controllers when relying on different lawful grounds for processing, ensuring due respect for the data subject's rights and freedoms.

The most appropriate basis depends on the purpose and relationship with the data subject. The public sector is likely to rely on the public task lawful basis, which requires the legal power to be laid down by law. This means that the processing of personal data is necessary for the performance of a task carried out in the public interest or in the exercise of official authority. For example, a local authority may process personal data to provide public services, such as waste collection, social care, or education.

The private sector, on the other hand, may use different lawful bases depending on the context and nature of their data processing activities. For instance, a private company may process personal data based on consent, which means that the individual has given clear and specific consent for a particular purpose. Alternatively, a private company may process personal data based on a contract, which means that the processing is necessary for the performance of a contract with the individual or to take steps at their request before entering into a contract. For example, an online retailer may process personal data to deliver goods or services that the individual has ordered.

Consequently, the difference between the public and private sector's lawful bases for processing personal data reflects their different roles, goals and purposes in the context of smart cities. The public sector aims to serve the public interest and meet local administrative obligations, while the private sector seeks to generate profit and fulfil contractual obligations.

### 3.4 Public-Private Partnerships and Concurrent Lawful Bases for Data Processing

Is it possible to apply multiple lawful bases to a single data processing activity? The Article 29 Working Party's Guidelines on Consent under Regulation 2016/679 clarify that one of the six legal grounds outlined in Article 6(1) GDPR must be identified prior to processing, which must be specific to the intended purpose. According to these guidelines (Sect. 5.2), relying on multiple lawful bases for a single purpose is not permitted. Once initiated, the lawful basis for processing cannot be altered, thus prohibiting controllers from toggling between different legal grounds.

Nevertheless, it could be argued that the GDPR permits the concurrent assignment of multiple lawful bases to a single processing purpose. This perspective stems from Article 6(1) GDPR, which deems data processing lawful provided it fulfills 'at least' one of the lawful bases enumerated within the same article. But what is the situation when joint controllers exist? According to the EDPB Guidelines 07/2020 on the concepts of controller and processor under the GDPR (page 4), each joint controller must ensure they have a legal basis for processing. Then, in footnote 56, the Guidelines mention that "[a]lthough the GDPR does not preclude joint controllers to use different legal basis for different processing operations they carry out, it is recommended to use, whenever possible, the same legal basis for a particular purpose." The Guidelines do not elaborate on the reasoning behind this recommendation, and the EDPB's position seems to reflect a practical approach rather than a strictly legal one.

Despite the above, the possibility that joint controllers might base data processing operations to fulfil the same purpose on various lawful grounds presents significant challenges. Firstly, it implies an overlap of diverse requirements - linked to the specific conditions that different lawful bases can impose and their feasibility in certain circumstances, for example, in processing special categories of data or when one of the controllers involved in the processing is subject to Directive 2016/680 which protects personal data in law enforcement activities (LED). In the same vein, data subjects' expectations and power imbalances are relevant when assessing the lawfulness of certain legal bases. It's worth noting here that Recital 43 GDPR states that a clear imbalance exists between the data subject and the controller when the controller is a public authority. This raises the question: Does this imbalance also exist in a scenario of joint controllership involving PPPs, where data processing relies on multiple lawful bases, and one of the joint controllers is a public authority? Finally, transparency concerns and the legal basis for data processing are crucial when addressing data subjects' rights. For example, when personal information is processed based on consent, data subjects can revoke this consent, requiring the data controller to cease processing the collected data. If a data subject exercises this right against a joint controller processing data based on public interest (e.g., a public authority), the situation becomes even more complex.

## 4    Concluding Remarks and the Way Forward

The exploration of lawful bases for data processing in PPPs contexts reveals a complex landscape. Joint controllers must address the interplay between public interest and individual rights, balancing the need for innovation and efficiency with the imperative of data protection and privacy. The public sector's inclination towards tasks carried out in the public interest contrasts with the private sector's contractual obligations and consent-based processing. This dichotomy underscores the distinct roles and responsibilities each sector bears in processing personal data, and it brings to the forefront the issue of whether different interests can be aligned, particularly regarding the potential for data to be monetised.

The challenges of applying multiple lawful bases to a single data processing activity highlight the need for clarity and consistency. The potential for power imbalances and the necessity for genuine choice in consent-based processing call for a cautious approach. Moreover, the requirement for transparency and accountability in informing data subjects about the purposes and lawful grounds for data processing cannot be overstated.

As we chart the course for smart cities and public-private partnerships, the need for clarity and stakeholder engagement stands out. Enhanced clarity on the lawful bases for data processing will empower all parties, especially data subjects. This clarity can be achieved through comprehensive guidelines and best practices tailored for joint controllers, and ensuring GDPR-compliant joint controllership agreements. However, it's also crucial to recognise specific scenarios where joint controllership is overly complex, making a controller-processor arrangement a more suitable choice.

# References

1. Richter, H.: Data access. In: Consumer Interests and Public Welfare, pp. 529–572. Nomos Verlagsgesellschaft mbH & Co. KG (2021)
2. Sadowski, J., Pasquale, F.: The spectrum of control: a social theory of the smart city. First Monday **20**(7) (2015)
3. Yovanof, G.S., Hazapis, G.N.: An architectural framework and enabling wireless technologies for digital cities & intelligent urban environments. Wirel. Pers. Commun. **49**, 445–463 (2009)
4. Foord, J.: The new boomtown? Creative city to Tech City in east London. Cities **33**, 51–60 (2013)
5. Batty, M., et al.: Smart cities of the future. Eur. Phys. J. Spec. Top. **214**, 481–518 (2012)
6. Anthopoulos, L., Fitsilis, P.: From digital to ubiquitous cities: defining a common architecture for urban development. In: 2010 Sixth International Conference on Intelligent Environments, pp. 301–306. IEEE (2010)
7. Sairamesh, J., Lee, A., Anania, L.: Information cities. Commun. ACM **47**, 29–31 (2004)
8. Anthopoulos, L.G., Reddick, C.G.: Smart city and smart government: synonymous or complementary? In: Proceedings of the 25th International Conference Companion on World Wide Web, pp. 351–355 (2016)
9. Ziosi, M., Hewitt, B., Juneja, P., Taddeo, M., Floridi, L.: Smart cities: mapping their ethical implications. SSRN Electron J. **10** (2022)
10. Rossi, U.: The variegated economics and the potential politics of the smart city. Territory Polit. Governance **4**(3), 337–353 (2016)
11. Voorwinden, A.: The privatised city: technology and public-private partnerships in the smart city. Law Innov. Technol. **13**(2), 439–463 (2021)
12. McNeill, D.: Global firms and smart technologies: IBM and the reduction of cities. Trans. Inst. Br. Geograph. **40**(4), 562–574 (2015)
13. Christofi, A., et al.: Smart city privacy: enhancing collaborative transparency in the regulatory ecosystem. In: 2019 CTTE-FITCE: Smart Cities & Information and Communication Technology (CTTE-FITCE), pp. 1–5. IEEE (2019)

14. Calzada, I.: Smart City Citizenship. Elsevier (2020)
15. Haarstad, H.: Who is driving the 'smart city' agenda? Assessing smartness as a governance strategy for cities in Europe. Serv. Green Econ. **1**, 99–5218 (2016)
16. European Commission: Green Paper on public-private partnerships and Community law on public contracts and concessions (2004)
17. Hood, C.: The "new public management" in the 1980s: variations on a theme. Account. Organ. Soc. **20**(2-3), 93–109 (1995)
18. Hodge, G.A., Greve, C., Boardman, A.E.: International Handbook on Public-Private Partnership. Edward Elgar Publishing (2010)
19. Forrer, J., Kee, J.E., Newcomer, K.E., Boyer, E.: Public–private partnerships and the public accountability question. Public Adm. Rev. **70**(3), 475–484 (2010)
20. Kitchin, R., Cardullo, P., Di Feliciantonio, C.: Citizenship, justice, and the right to the smart city. Right Smart City, 1–24 (2019)
21. Goldstein, B.T., Mele, C.: Governance within public–private partnerships and the politics of urban development. Space Polity **20**(2) 194–211 (2016)
22. Gaffney, C., Robertson, C.: Smarter than smart: Rio de Janeiro's flawed emergence as a smart city. J. Urban Technol. **25**(3), 47–64 (2018)
23. Sanchez, M.G.: Joint controllership in data protection: what the CJEU has stated so far. Edinburgh Student L. Rev. **4**(86) (2020)
24. G20 Global Smart Cities Alliance: Primer for smart city public-private collaborations. EWorld Econ. Forum. (2022)
25. Judgment in Wirtschaftsakademie, C-210/16, ECLI:EU:C:2018:388, paragraph 38
26. Fashion ID, C-40/17, ECLI:EU:2018:1039, paragraph 74

# Privacy Promise Vs. Tracking Reality in Pay-or-Tracking Walls

Timo Müller-Tribbensee[✉][iD]

Goethe University Frankfurt, Frankfurt am Main, Germany
mueller-tribbensee@wiwi.uni-frankfurt.de

**Abstract.** European websites increasingly adopt pay-or-tracking walls, sometimes known as "consent or pay models," "cookie paywalls," or "pay-or-okay walls." These walls require users to pay a fee or consent to be tracked in exchange for website access. However, initial evidence suggests that websites might continue to track users even when they pay the fee, constituting user deception. This paper comprehensively assesses whether websites employing pay-or-tracking walls keep their privacy promise to paying users as stated on the pay-or-tracking wall and safeguard their privacy. Data collection and analysis from 341 websites show that while websites reduce tracking for paying users, 32.9% of the websites fail to uphold the privacy promise declared on their pay-or-tracking wall. 80% of these websites could meet their privacy commitments by removing just one or two trackers. Notably, a group of websites offering a joint subscription allowing access to all participating websites better keeps their privacy promises than others, likely due to the implementation of an ongoing control mechanism that regularly detects tracker usage. The results show that implementing tracking-free websites remains challenging and might require continuous efforts.

**Keywords:** Privacy · Consent · Tracking · Pay-or-Tracking Wall

## 1 Introduction

Due to European privacy laws, particularly the European ePrivacy Directive (ePD) and the General Data Protection Regulation (GDPR), websites must obtain users' consent for tracking. To comply, websites commonly display consent notices, asking users whether they accept or decline tracking, such as for advertising purposes. While refusing tracking has traditionally incurred no monetary cost for users, European websites increasingly adopt pay-or-tracking walls, sometimes known as "consent or pay models," "cookie paywalls," "pay-or-okay walls," or "accept-or-pay cookie banners" [21, 22, 24]. These pay-or-tracking walls require users to either pay a fee or consent to be tracked in exchange for access to the website. While the latter ("tracking option") involves no payment but the user's consent for tracking, the former ("pay option") costs money, requires no consent for tracking, and comes with a text that highlights its privacy advantage, such as "without tracking," which we refer to as the "privacy promise."

M. Jensen et al. (Eds.): APF 2024, LNCS 14831, pp. 168–188, 2024.
https://doi.org/10.1007/978-3-031-68024-3_9

However, initial evidence suggests that websites might continue to track users even when they purchase the pay option for accessing the website without providing consent for tracking [21, 22]. Such reports raise questions about how websites using pay-or-tracking walls fulfill their stated privacy promise for paying users and whether they align with the actual tracking practice.

If websites break their privacy promises by tracking users even after receiving payment, it will not only deceive users but might also breach legal standards regarding consent under Article 5(3) of the ePD, unless such tracking is deemed strictly necessary for the operation of the website. Conversely, if the allegations are false and the pay option's implementations are privacy-preserving, it would show that pay-or-tracking walls are better than their reputation and an adequate alternative to protect users' online privacy while supporting websites' monetization.

This paper assesses whether websites employing a pay-or-tracking wall keep their privacy promises and offer pay options that safeguard user privacy. It addresses the following research questions (RQ):

- What do websites promise users who purchase the pay option? (RQ1)
- What tracking do websites implement in pay-or-tracking walls' choice alternatives? (RQ2)
- To what extent do websites keep their privacy promises? (RQ3)

## 2 Overview of Tracking

### 2.1 Value, Privacy Risk, and Technologies of Trackers

Tracking is often defined as collecting users' online activities over time (e.g., [15, 20]). It can be particularly harmful to privacy when the data recipients collect a relevant portion of the user's online activities across multiple websites [15], a practice referred to as cross-site tracking. This risk is particularly present when websites integrate widespread third-party services operating as a tracker. The latter enables websites to embed relevant functionalities, such as advertising (e.g., DoubleClick to manage ad sales), site analytics (e.g., Google Analytics to analyze website traffic), or social media (e.g., Facebook Social Plugins to integrate a widget). However, the integration may also enable the tracker to collect the user's online activities across multiple websites, for instance, by storing a third-party cookie on the user's device that allows re-identifying the user on different websites. While third-party cookies are a well-known method, trackers employ a range of technologies, also including fingerprinting or tracking pixels (e.g., [3, 25]).

### 2.2 Tracker Vs. Essential Third-Party Service

The GDPR requires that any processing of personal data must be carried out by data controllers based on a legal basis (as listed in Article 6), such as consent or legitimate interest. Additionally, tracking is further restricted by Article

5(3) of the ePD, which mandates consent when operations involve storing and gaining access to information on the user's device, as in the case of trackers, and regardless of the specific technology used [7]. Under the GDPR, Article 4(11) and Article 7 define consent as a voluntary and explicit choice (among other requirements). Consequently, websites must ask users for consent to track them and embed trackers.

However, websites' technical functionality frequently depends on third-party services, and these essential third-party services often store and retrieve information from the user's device. Accordingly, Article 5(3) of the ePD provides exceptions to the consent requirement, which can apply when being deemed as "strictly necessary" for the website's operation. Legal authorities interpret these exceptions rather narrowly and typically refer to purely technical necessity (e.g., [4]).

Examples of essential third-party services that websites deem technically necessary involve, for instance, tag managers. Websites need tag managers to centrally manage scripts, integrate dynamic elements, and streamline privacy compliance efforts. Another example of essential third-party services are site analytics tools, which are crucial for generating aggregate performance statistics, detecting navigation issues, or optimizing technical performance.

However, since these essential services, including tag managers and site analytics tools, are often maintained by third parties, they may transmit user information, such as IP addresses, or store third-party cookies, potentially enabling the third-party service to track the user (even across websites). Therefore, websites should ensure that essential third-party services are designed and configured to minimize the risk of being misused for tracking purposes beyond the essential technical purpose. Further, websites rely on third-party services to handle the transmitted information responsibly. Due to these risks, companies that provide anti-tracking protection, such as Ghostery, often classify essential third-party services as trackers [13].

Moreover, in the case of third-party site analytics services, legal authorities may consider them essential but only under strict conditions. For instance, the French Data Protection Authority mandates that such third-party site analytics services must produce anonymous statistical data only, not use a global user identifier across different websites, and limit its sole purpose strictly to site analytics to be exempt from consent [10].

While essential third-party services can pose a risk to user privacy, their technical necessity and legal recognition for being exempt from consent suggest a distinction between trackers and essential third-party services, which we consider in the analysis of this paper.

# 3   Related Work

## 3.1   Discrepancy Between Promise and Reality for Cost-Free Decisions to Refuse Consent or Opt Out

Many privacy laws, such as the European privacy regime, and industry initiatives, such as Apple's App Tracking Transparency framework, aim to mitigate privacy risks by empowering users via explicit consent (opt in) or enabling users to object to (default) tracking activities (opt out). Thus, in addition to the tracking option, websites and apps typically provide a second alternative, allowing users to access a website or app cost-free without tracking.

However, previous research has revealed discrepancies between privacy promises and actual tracking practices in the context of such cost-free and supposedly tracking-free choices. For instance, Matte et al. [19] document instances of websites storing positive consent even when users refuse consent. Further, Bouhoula et al. [1] find that websites often continue tracking despite users choosing to refuse. Similarly, Bui et al. [2] show that embedded third-party services on websites may continue tracking despite users choosing the cost-free opt-out option, contradicting the privacy promise not to track opt-out users. Similar discrepancies have been observed for apps [5], with multiple studies highlighting that privacy labels for apps are often misleading and diverge from the tracking reality (e.g., [16,17]). Further studies aim to explain such discrepancies. For instance, Utz et al. [26] report that web developers often lack awareness regarding data collection of third-party services and that privacy plays a minor role in their implementation decisions.

## 3.2   Potential Discrepancy Between Promise and Reality in Pay-or-Tracking Walls

Providing a tracking-free alternative and keeping the privacy promise is even more critical in the context of pay-or-tracking walls. The pay option offered by European websites represents the legally required possibility to refuse consent for tracking under European privacy laws, and it is a service the user pays for.

Three studies have examined pay-or-tracking walls, with two indicating that paying users may still be subject to tracking [21,22], while a third study reports the opposite finding [24]. First, these studies do not address whether websites transparently include certain tracking activities within the pay option as part of their pay option's privacy promise. Secondly, the three previous studies exhibit limitations that prevent a conclusive determination of whether paying users are being tracked, involving no measurement of embedded trackers [21], no distinction between trackers and essential third-party services [22], and no measurement of tracking beyond third-party cookies [24]. Thus, this study aims to comprehensively assess whether websites keep their privacy promises and offer pay options that safeguard user privacy. We elaborate on the details of each of the three previous studies on pay-or-tracking walls and this study's contribution in what follows.

First, Morel et al. [21] highlight the growing trend of European websites charging users for data protection, identifying 431 websites employing a pay-or-tracking wall via an automated approach. They further examine the websites' implementation of the Transparency and Consent Framework (TCF), an industry standard for websites to streamline consent decisions and disseminate them with the website's implemented third-party services. They find that even if users choose the pay option and refuse to consent, they might be tracked for inappropriate purposes, such as advertising, which do not fall under the consent exemption of Article 5(3) of the ePD as outlined by the authors. However, the analysis focuses on the disseminated consent string via the TCF but includes no measurement of embedded trackers on the websites. Thus, it is unclear whether only the TCF consent string is unlawful or whether users are actually tracked.

Second, Müller-Tribbensee et al. [22] analyze the popularity of pay-or-tracking walls, design, price, and user reactions among top European websites. Regarding tracking activities, they examine 26 top European websites using a pay-or-tracking wall and discover the presence of trackers on most websites even after purchasing the pay option. However, it is unclear whether some of the identified tracking activities are essential third-party services deemed technically necessary for the website's operation.

Third, Rasaii et al. [24] developed an automated approach to detect pay-or-tracking walls and analyze tracking cookies in the pay option of 219 websites. They find no evidence for tracking after purchasing the pay option. Their analysis focuses on a group of websites offering a joint subscription for the pay option, allowing access to all participating websites. The group is managed by an independent provider, which the authors refer to as a subscription management platform. Such websites participating in the group that offer a joint subscription potentially implement a different tracking regime than other websites, as the group may need to implement the same tracking policy. More importantly, the study focuses on tracking via third-party cookies. Thus, whether the websites continue tracking paying users via different technologies, such as tracking pixels, remains unclear.

Motivated by the insights and mixed findings from prior studies, this paper's contribution entails examining the alignment between the stated privacy promises and the actual tracking practice for pay-or-tracking walls. Further, this study covers various tracking technologies (e.g., third-party cookies, pixels) across different website types (e.g., those participating in a group offering a joint subscription or not). It also considers the distinction between trackers and essential third-party services. Lastly, this paper analyzes the potential discrepancy between what users pay for and receive.

# 4    Setup of Empirical Study

## 4.1    Sample

This study uses the website list of pay-or-tracking walls provided by Morel et al. [21], including a country and website industry classification[1] We focus on German-speaking countries, i.e., Austria, Germany, and Switzerland, representing 341 websites and 79% of all identified websites using pay-or-tracking walls. We visited the selected websites between October 6 and November 26, 2023, from an IP address in Germany using the Google Chrome browser (version 117.0.5938.150). We purchased the pay option for each website and collected data about their stated privacy promise and actual tracking activities.

## 4.2    Collecting Privacy Promises

As shown in Fig. 1, a pay-or-tracking wall typically gives users the choice between the pay and tracking option upon their first website visit. While the latter asks for consent and informs the user about the tracking purposes, the pay option requires no consent and may include additional benefits, such as reduced advertising or access to premium content [22]. Further, the pay option is presented along with a privacy promise, which involves a text highlighting its privacy advantage, such as "without tracking."

We retrieve the privacy promise associated with the website's pay option using a manual approach, allowing us to reliably analyze the relevant texts from the pay-or-tracking wall's first layer (i.e., what users see upon the first website visit). For each website visit, we use the Google Chrome browser in Incognito mode, imitating a first-time visitor without browsing history. We refer to that configuration as a "clean" browser instance. We further configured the browser's language settings so that websites would show their content preferably in English. This led to websites displaying their pay-or-tracking wall, if available, in English and otherwise in German. We document each pay-or-tracking wall with screenshots and translate German privacy promises into English.

## 4.3    Measuring Tracking Reality

To detect the presence of trackers during our website visits, we use the browser extension Ghostery Insights [12], which is maintained by Ghostery, a well-known privacy protection company that offers anti-tracking and ad-blocking software. The browser extension exposes trackers during a website visit by logging network traffic and comparing embedded third-party requests on the website with their

---

[1] Morel et al. [21] classify websites' country by using geographical information associated with the registrant of a domain and websites' industry by using a commercial tool analyzing websites' content. Their website list is available in this Google Sheet. https://docs.google.com/spreadsheets/d/1UBiIaH5LAf04IlDsnf7zm68b_csA4_KiJDZyaZcLdBc/edit#gid=0.

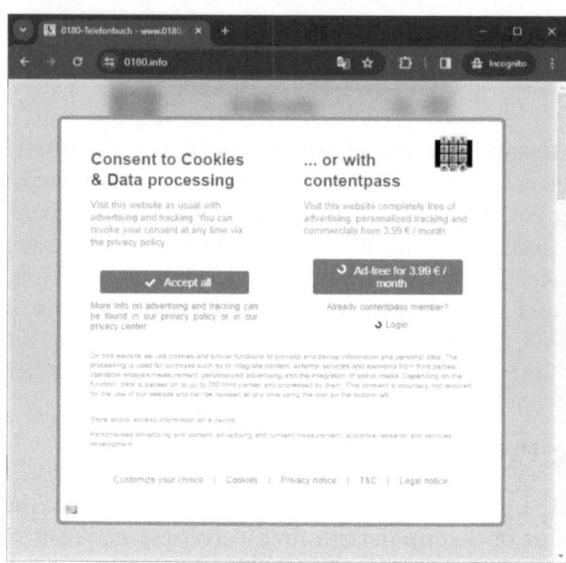

**Fig. 1.** Exemplary pay-or-tracking wall of the website "0180.info". *Notes*: Screenshot taken on October 6, 2023. Pay option ("... or with contentpass") displayed on the right half, promising website access "completely free of personalized tracking" and without advertisements.

database of known tracking domains. It detects trackers using various technologies, such as third-party cookies or tracking pixels.

The browser extension allows for extracting two files for each visited webpage containing information on the detected trackers. The first file offers detailed information, such as the specific tracker domain linked to the tracker, while the second file provides a summary, including tracker names and tracker categories. The tracker categories are closely related to the tracker's functionality for the website, such as advertising, site analytics, social media, or essential [13], and have been used in previous research settings on online tracking (e.g., [18,23]).

The tracker categories are relevant to our setting for two purposes. First, it allows us to compare whether the privacy promise aligns with the actual tracking practice, particularly in pay options involving a privacy promise excluding specific tracker categories. For instance, if a website promises a pay option "without ad tracking," we can measure the presence of trackers from the advertising category to compare the promise with the tracking reality.

Secondly, the tracker categories enable us to distinguish between trackers and essential third-party services, with the latter being utilized by websites for technical purposes, not for tracking user behavior. To identify and exclude such essential third-party services from our analysis, we refer to Ghostery's "essential" category, described as *"Site requests that may be critical to website functionality, such as tag managers and privacy notices"* [13].

Additionally, we acknowledge the technical necessity for websites to perform site analytics, provided these third-party services are privacy-friendly. We refer to the list published by the French Data Protection Authority [10], which includes third-party site analytics services that can be configured to fall within the scope of the consent exemption of the ePD and produce anonymous statistical data only. Moreover, we exclude the privacy-friendly third-party site analytics service of VG Wort, a German non-profit organization distributing royalties on behalf of authors and publishers. The latter's sole purpose is to derive the aggregate number of page views for an article, and it qualifies for the consent exemption of the ePD as outlined by the provider VG Wort [27].

In summary, we exclude from our list of detected trackers 1) those identified as part of Ghostery's "essential" category and 2) privacy-friendly third-party site analytics services, including VG Wort and those listed by the French Data Protection Authority [10]. Specifically, the latter list includes the third-party site analytics services of AT Internet, Etracker, Matomo, and Piwik Pro, which were identified during our data collection. However, essential third-party services still represent a risk to users being tracked due to a wrong configuration of the third-party service. Thus, we also conduct robustness checks by including essential third-party services in an additional analysis, particularly those classified as essential by Ghostery.

Regarding the procedure, we measure each website's tracking activities 1) before making the choice between the pay and tracking option, 2) after selecting the tracking option, and 3) after purchasing the pay option, allowing us to evaluate the pay option's privacy advantage compared to the other scenarios. Further, we manually visit each website to reliably ensure the selection of each alternative option and the proper login to the pay option. Using a clean browser instance, we first detect tracking activities before making the choice between the pay and tracking options. Next, we select the tracking option and detect tracker usage on the website's start page. Since tracker usage on subpages like news articles may differ from the start page, we further analyze trackers on two additional subpages. We picked these subpages randomly among those featured on the website's start page. Third, we again open a clean browser instance and purchase the pay option. We then log in to the pay option and measure the tracking activities on the website's start page and the same set of two subpages. After collecting data on websites' tracking activities, we reviewed the list of detected trackers and removed the previously described essential third-party services for our main analysis.

Despite the advantages of capturing tracking across various tracking technologies and distinguishing between trackers and essential third-party services, the presented approach has some limitations. First, we may not detect all trackers, for instance, due to the collection procedure only involving a subset of the website's subpages or the browser extension Ghostery Insights relying on lists of known tracking domains. Although the latter is a well-known approach in the research community (e.g., [6,14]), tracker companies may circumvent detection, for instance, by registering new domains that are not yet included on these lists [9]. Second, although embedding a tracker into a website enables the company behind the tracker to collect and process data, we do not always know whether the company actually stores and uses such personal data. However, embedding trackers increases users' vulnerability to potential misuse or unauthorized access, thus harming privacy.

## 5    Results of Empirical Study

### 5.1    Sample Description

We refined the dataset by excluding 49 websites from the initial list of 341 German-catering websites. Out of the 49 excluded websites, 43 redirect users to other websites in our sample, one website did not use a pay-or-tracking wall, and another did not cater to a German-speaking audience - the other four exclusions comprised websites with technical issues during the purchase of the pay option. Despite efforts to resolve these problems by contacting the websites, they remained unresolved or unanswered. The final sample consists of 292 websites.

Additionally, we faced technical problems after purchasing the pay option on several websites, which we outline in more detail in the Appendix A in Table 2. Some issues involved websites displaying the banner for the pay-or-tracking wall even after logging in to the pay option. However, most websites resolved this problem after we contacted them. Moreover, some websites lacked a dedicated login button for the pay option on their pay-or-tracking wall. Notably, three websites suggested clicking the "accept tracking" button as a workaround to either remove the persistent banner of the pay-or-tracking wall or gain access to the login button for the pay option on the start page.

As depicted in the left upper panel of Fig. 2, the final sample of 292 websites consists of websites from Germany (92.1%), Austria (7.2%), and Switzerland (0.7%). Regarding the website industry (depicted in the right panel of Fig. 2), the majority concerned News (23.3%), followed by Business (16.8%), Computers and Technology (14.4%), and a variety of other industries (45.5% in total and depicted in more detail in Fig. 2).

Moreover, we observe two types of pay-or-tracking wall offerings. As portrayed in the lower left panel of Fig. 2, the first one comprises websites selling their "Own Pay Option," allowing users to access only one or a few websites of a single publisher (36.6%). The second type involves a group of websites offering a joint subscription for the pay option, allowing access to all participating websites. The group is managed by an independent provider known as Contentpass, which charges a collective fee for the pay option and enables users to access multiple websites from various publishers. We refer to the latter group as "Multi-Website Provider" (63.4%).

### 5.2    RQ1: What Do Websites Promise Users Who Purchase the Pay Option?

As illustrated in Fig. 4, we identified 17 distinct formulations of privacy promises for the pay option among the sample of 292 websites. The most prominent formu-

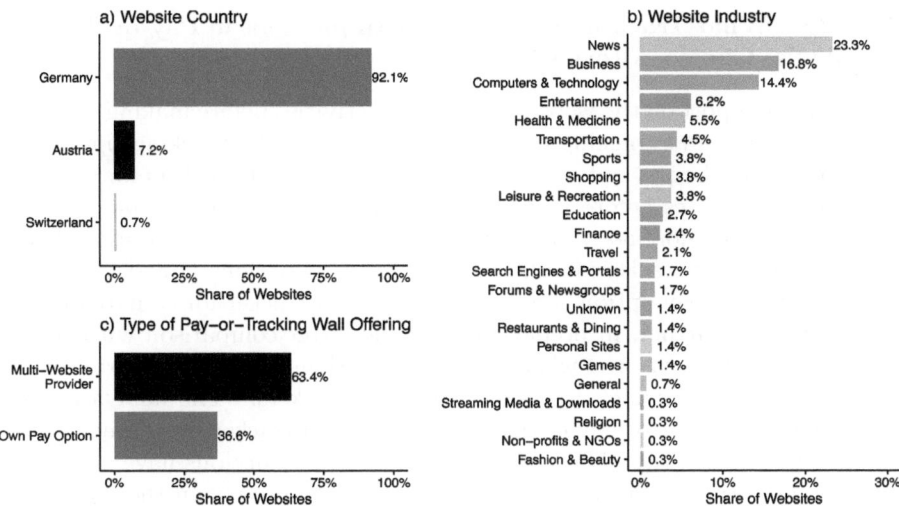

**Fig. 2.** Sample description by website country, website industry, and type of pay-or-tracking wall offering. *Notes*: N = 292 websites. Website industry and country classification adopted from Morel et al. [21]. "Multi-Website Provider" refers to websites participating in the group of websites offering a joint subscription for the pay option. "Own Pay Option" refers to websites selling their own pay option, allowing users to access only the website(s) of a single publisher. Percentages may not sum up to exactly 100% due to rounding.

lations include "Without Personalized Tracking" (29.1%), "Without Ad Tracking" (22.6%), "Completely Free of Personalized Tracking" (20.2%), "Without Ad Tracking and Marketing Cookies" (9.9%), and "Largely Free of Tracking" (7.5%). Various less prominent formulations comprise 10.7% in total, as depicted in more detail in Fig. 4.

In terms of clarity, the promise "Largely Free of Tracking" stands out as a vague statement. Additionally, some formulations, such as "Without Personalized Tracking," include the term "personalized," the meaning of which remains unclear. Regarding content, most privacy promises involve a general commitment not to track users. In contrast, the remaining privacy promises only exclude tracking activities related to advertising, representing a weaker privacy promise than not tracking users.

We utilized the difference among the promises' content to manually categorize the 17 distinct formulations of privacy promises into the two groups: those with a general commitment of "No Tracking" and those with a weaker commitment of "No Ad Tracking." As portrayed in Fig. 4, the "No Tracking" category comprises 65.4% of websites, while the "No Ad Tracking" category accounts for 34.6%. Although we classified the formulation "Largely Free of Tracking" under the "No Tracking" category, we will also analyze this vague statement separately in our subsequent analysis when comparing the privacy promises to the tracking reality.

## 5.3    RQ2: What Tracking Do Websites Implement in Pay-or-Tracking Walls' Choice Alternatives?

In Table 1, we report on the websites' tracking activities before making the choice between the pay and tracking option, after selecting the tracking option, and after purchasing the pay option. Firstly, we do not detect trackers for a share of 76.4% of the websites before making the choice. The remaining share of 23.6% tracks users even before users choose between the pay or tracking option. Unsurprisingly, we detect trackers in all websites' tracking options (100.0%).

Regarding the pay option, we do not detect any tracker for a share of 62.7% of the websites. Further, in preparation for the latter comparison with the privacy promises, we measured whether the pay options include trackers from the advertising category, which we refer to as "ad trackers." The analysis reveals that a share of 12.0 % of pay options includes no ad tracker but other trackers, such as for site analytics. The remaining 25.3% of pay options have ad trackers (and possibly other trackers). While trackers can be present in the pay option, their number is lower compared to the tracking option. The number of trackers typically ranges between 0 and 3 in the pay option, referring to the lower 5% and the upper 95% percentile. In comparison, the tracking option typically includes between 6 and 91 trackers.

**Table 1.** Descriptive statistics of tracking activities per option.

| Metric | Before Making the Choice | Tracking Option | Pay Option |
|---|---|---|---|
| *Number of Distinct Trackers* | | | |
|   Minimum | 0 | 2 | 0 |
|   5% Percentile | 0 | 6 | 0 |
|   Median | 0 | 56 | 0 |
|   Mean | 1 | 48 | 1 |
|   95% Percentile | 2 | 91 | 3 |
|   Maximum | 66 | 96 | 21 |
| *Share of* | | | |
|   No Tracker Detected | 76.4% | 0.0% | 62.7% |
|   No Ad Tracker But Other Tracker Detected | 4.1% | 0.0% | 12.0% |
|   Ad Tracker Detected (& Possibly Other Tracker) | 19.5% | 100.0% | 25.3% |

*Notes*: N = 292 websites. "Ad Tracker" refers to a tracker from the advertising category. Excluded trackers: Trackers from "essential" category, AT Internet, Etracker, Matomo, Piwik Pro, and VG Wort. Percentages may not sum up to exactly 100% due to rounding.

In the Appendix A in Fig. 5, we further analyze the prevalence of each tracker category in each option (i.e., before making the choice, pay option, tracking option) in more detail, including essential third-party services not considered as trackers (e.g., tag managers). If the website's pay option and before making a choice include tracking, the detected trackers are typically part of the advertising or site analytics category. The tracking option may include additional tracker categories, such as customer interaction or social media.

Moreover, we analyze those websites embedding tracking in the pay option and before making the choice in more detail. As portrayed in the right panel of Fig. 3, most websites embedding tracking in the pay option include one (55.0%) or two (24.8%) trackers. Thus, for around 80% of the websites still embedding tracking, a tracking-free pay option could be achieved by removing only one or two trackers. Similarly, the left panel of Fig. 3 provides an overview of tracking before making the choice, revealing that most websites embedding tracking incorporate one (55.1%) or two (23.2%) trackers.

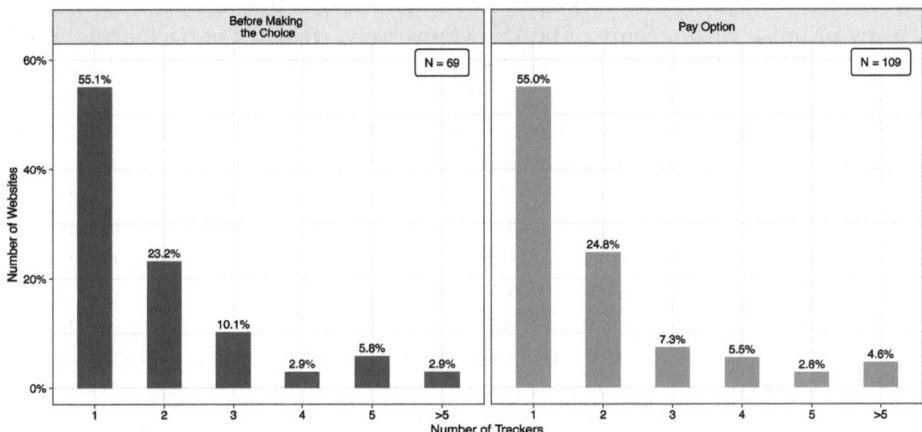

**Fig. 3.** Number of trackers for websites still embedding tracking (before making the choice and pay option). *Notes*: N = 69 websites (before making the choice), N = 109 websites (pay option). "Before Making the Choice" based on tracking measured on the starting page. "Pay Option" based on tracking measured on the starting page and two additional subpages. Excluded trackers: trackers from "essential" category, AT Internet, Etracker, Matomo, Piwik Pro, and VG Wort.

### 5.4   RQ3: To What Extent Do Websites Keep Their Privacy Promises?

When assessing the extent to which websites keep their privacy promises, we assume that the website should not involve any tracker to keep the promise of the "No Tracking" category. Similarly, promises of the "No Ad Tracking" category require websites not to embed ad trackers.

As depicted in Fig. 4, the analysis reveals that a share of 32.9% of the websites do not keep their pay option's privacy promise. Despite websites breaching their privacy promise, some websites provide more tracking protection than promised. For instance, those websites that only commit to "No Ad Tracking" but, in fact, do not track the user at all. Further, we specifically examine websites that use the vague privacy promise "Largely Free of Tracking." Our findings show that these websites do not embed trackers, effectively making them tracking-free.

Next to our main analysis revealing that a share of 32.9% of the websites do not keep their pay option's privacy promise, we also conduct a robustness check by including essential third-party services, particularly those classified as essential by Ghostery. As depicted in the Appendix A in Fig. 6, the share of websites not keeping their privacy promise would increase to 36.3% when treating third-party services from Ghostery's "essential" category as trackers.

We conduct further analysis to identify heterogeneities between websites. While the websites' industry and country do not indicate clear differences, the two types of pay-or-tracking wall offerings do. The websites that sell their own pay option, each providing users only access to one or a few websites, break their privacy promise with a share of 60.7%. Conversely, websites participating in the multi-website provider offering a joint pay option allowing access to multiple websites break their privacy promise with a significantly lower share of 16.8%.

We contacted Contentpass, the company behind the multi-website provider, to learn more about the possible reasons why participating websites better safeguard users' privacy than others. The company stated that this is likely due to their ongoing external control of website tracking activities, which include running daily crawls to detect trackers on participating websites and requesting them to remove trackers in their pay options.

In response to why websites keep embedding trackers in the pay option, Contentpass reports that website developers and editors often unintentionally integrate trackers due to their lack of awareness of the trackers' data collection practices. Moreover, many websites have evolved organically and were not originally designed with the explicit objective of prioritizing user privacy.

# 6   Summary of Results and Implications

## 6.1   Summary of Results

This paper analyzes whether websites using a pay-or-tracking wall keep their privacy promises. Data collection preceding the analysis shows that some websites have implemented pay-or-tracking walls inadequately. While the implementation problems only concern a minority of websites, we could not purchase the pay option on four websites despite contacting them. Moreover, we encountered several problems after purchasing the pay option on some websites, such as no dedicated login button for the pay option on the banner displaying the pay-or-tracking wall.

The data analysis of 292 websites reveals that websites use 17 distinct formulations for their pay option's privacy promise. The content of these promises falls into two categories: "No Tracking" and the weaker promise of "No Ad Tracking." Regarding tracking, we detect trackers in 37.3% of the websites' pay options, while 62.7 % are tracker-free, thus truly safeguarding user privacy. The websites that still integrate tracking in the pay option typically embed one (55.0%) or two (24.8%) trackers. Moreover, we find that 23.6% of the websites track users even before making the choice for the pay or tracking options. When comparing the pay option's privacy promise with the actual tracking reality, we find that

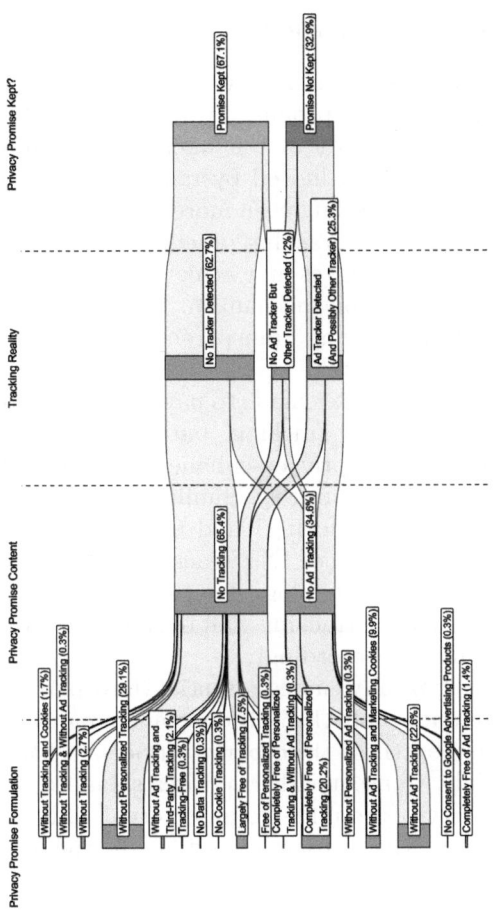

**Fig. 4.** Privacy promise vs. tracking reality. *Notes*: N = 292 websites. "Privacy Promise Formulation" refers to the collected promises associated with the pay option. "Privacy Promise Content" describes the categorized content of the privacy promises. "Tracking Reality" shows the actual tracking activities detected after selecting the pay option (excluded trackers: trackers from "essential" category, AT Internet, Etracker, Matomo, Piwik Pro, and VG Wort). "Privacy Promise Kept?" shows whether the "Privacy Promise Content" aligns with the "Tracking Reality". Percentages may not sum up to exactly 100% due to rounding.

32.9% of websites do not keep the privacy promise associated with their pay option.

The group of websites offering a joint subscription for the pay option (referred to as "Multi-Website Provider"), allowing paying users access to all participating websites, noticeably keeps their privacy promise more often than websites selling their own pay option. This difference may be explained by the ongoing efforts of the group to detect tracker usage on participating websites.

## 6.2   Implications and Conclusions for Websites

We observe that 37.3% of the websites' pay options are not tracking-free, and 32.9% do not keep their privacy promise, so websites should review their tracking practices critically[2]. For 80% of the websites still embedding tracking, a tracking-free version could be achieved by removing only one or two trackers. Conversely, the websites embedding even more trackers in the pay option should comprehensively overhaul their tracking practices. As shown by the example of the multi-website provider, websites may achieve a better offering by implementing an ongoing tracker detection mechanism. Furthermore, making the results and actions from such monitoring transparent to users and regulators would enhance trust and transparency.

Regarding the privacy promises made to paying users, websites that currently use the weak promise of "No Ad Tracking" but do not track paying users should consider changing to the more privacy-enhancing promise of "No Tracking" to describe their offering more accurately. Similarly, websites that use the vague promise of "Largely Free of Tracking" and do not track paying users should consider changing to the promise of "No Tracking." While consent management platforms often provide the technology for the banner, websites can typically customize the text on a pay-or-tracking wall to suit their specific needs, allowing them to make these changes accordingly.

Moreover, websites typically embed essential third-party services, such as tag managers or site analytics tools that produce anonymous statistical data only. However, websites rely on these third-party services to handle the transmitted information responsibly. Consequently, websites should consider requiring such essential third-party services to undergo an external audit of their source codes to demonstrate they are privacy-friendly and handle the data responsibly.

## 6.3   Implications and Conclusions for Users

Most importantly, 37.3% of websites' pay options are not tracking-free, and 32.9% do not keep their privacy promise. Thus, users choosing the pay option risk wasting their money. Users who want to safeguard their privacy may need to continue using additional (and often cost-free) privacy protection tools, such as browser extensions blocking trackers. Additionally, 34.6% of the pay options

---

[2] We shared the findings with the multi-website provider, representing 185 websites in the sample of this study. Additionally, we presented the findings at an industry association meeting of the German Association for the Digital Economy (BVDW), which prompted over 24 websites to request their individual results (as of the time of writing this paper). We encourage any other interested websites to contact the author for their results from this study.

come with the weak privacy promise of "No Ad Tracking" instead of "No Tracking." Consequently, users considering a pay option should thoroughly examine the website's offering.

## 6.4  Implications and Conclusions for Regulators

Our findings reveal that 32.9% of websites fail to uphold their privacy promise to paying users, and 37.3% of websites' pay options are not tracking-free. These results show that implementing tracking-free websites remains challenging. Regulators may consider supporting (or even mandating) websites to implement measures to safeguard privacy more continuously, such as an ongoing tracker detection mechanism. Such measures may help improve user privacy and foster a fair, competitive landscape among websites. The latter becomes particularly relevant for regulators to maintain an equal level playing field between websites devoting more resources to ensure user privacy and those that do not.

Moreover, users interested in purchasing a pay option cannot be sure whether a website's pay option is entirely tracking-free. As tracker detection is complex, users may benefit from a standardized privacy label that externally validates whether a pay option is privacy-friendly. Such a privacy label could also include an open-access repository, such as a website, that transparently publishes the results of the (continuous) privacy checks. Regulators should consider initiating or supporting such initiatives to improve transparency and user trust. Some regulators have already taken first steps to increase transparency for users, such as the French Data Protection Authority, which offers a browser extension enabling users to identify cookies stored on their browser [11].

Lastly, the adoption of pay-or-tracking walls is fairly advanced in some countries, such as Austria and Germany. Despite the latest discussions on the general compliance of pay-or-tracking walls under European privacy laws [8], this trend may expand to other European countries with similar regulatory environments. Thus, other national regulators should prepare for the privacy challenges that arise with the advent of pay-or-tracking walls.

**Acknowledgments.** This study received financial support from the European Research Council (ERC) under the European Union's Horizon 2020 Research and Innovation Program (Grant Agreement No 833714). I would also like to thank Bernd Skiera, the participants of the Marketing Department's Doctoral Colloquium at Goethe University Frankfurt, and the anonymous reviewers for their valuable feedback and suggestions.

**Disclosure of Interests.** The author has no competing interests to declare that are relevant to the content of this article.

# A   Appendix

**Table 2.** Technical problems after purchasing the pay option.

| Problem | Website's Solution | Number of Websites | Share of Websites |
|---|---|---|---|
| The access to the pay option worked only one or two days after purchasing it | N/A | 5 | 1.7% |
| The pay option and its login are not visible on the first layer but only on further layers of the banner displaying the pay-or-tracking wall | N/A | 1 | 0.3% |
| After logging into the pay option, the pay-or-tracking wall banner returns | a) Resolved after contacting the website | 11 | 3.8% |
| | b) In an email providing the order confirmation for the pay option, the website refers to this problem and recommends users to click the "accept tracking" button to remove the banner | 1 | 0.3% |
| The banner displaying the pay-or-tracking wall has no dedicated login button for the pay option | After contacting the websites, they recommended clicking the "accept tracking" button and logging in on the start page | 2 | 0.7% |
| The pay option is not accessible via an account. Instead, the website stores a cookie in the browser. However, the website provides no way to recreate the cookie once deleted | N/A (Contacting the website and asking for a solution remained unanswered.) | 1 | 0.3% |

*Notes*: N = 292 websites

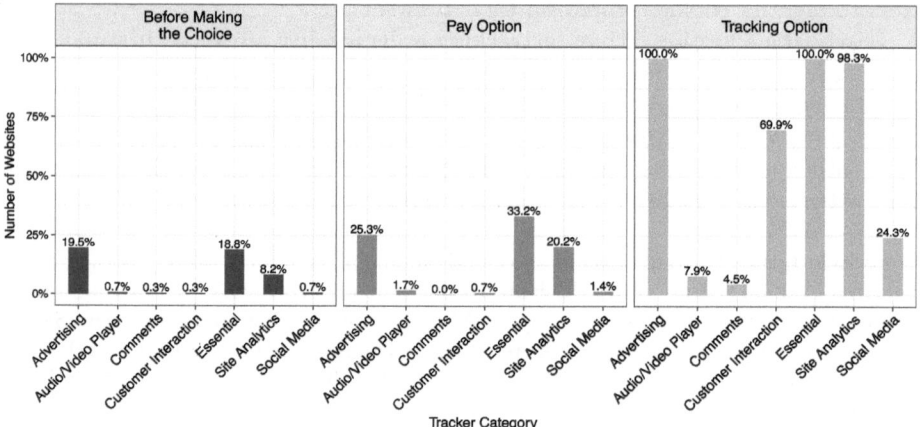

**Fig. 5.** Prevalence of tracker categories and essential third-party services per option. *Notes*: N = 292 websites. Each bar shows the number of websites that include trackers from the category. "Before Making the Choice" based on tracking measured on the starting page. "Pay Option" and "Tracking Option" based on tracking measured on the starting page and two additional subpages. Essential third-party services labelled as "Essential". Excluded trackers: AT Internet, Etracker, Matomo, Piwik Pro, and VG Wort.

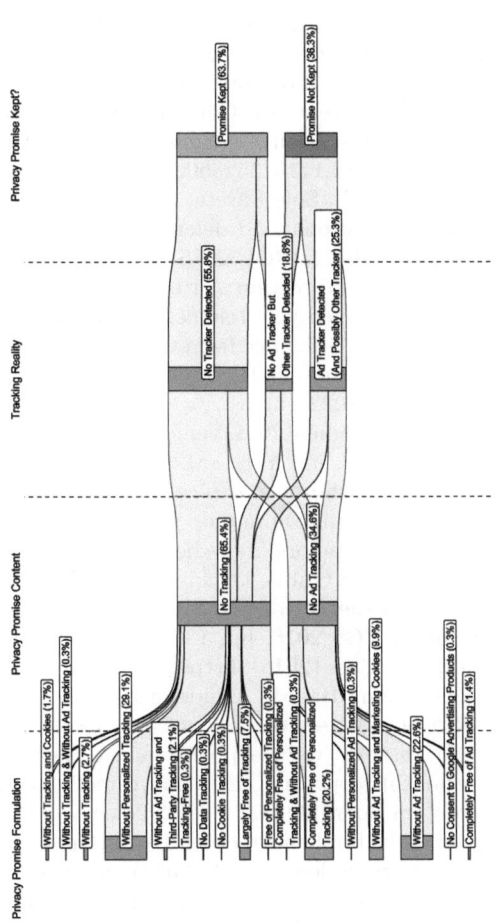

**Fig. 6.** Privacy promise vs. tracking reality (treating the "essential" category as tracker). *Notes*: N = 292 websites. Excluded trackers: AT Internet, Etracker, Matomo, Piwik Pro, and VG Wort. Percentages may not sum up to exactly 100% due to rounding. The figure presents a robustness check for the results of the main analysis, depicted in Fig. 4. In contrast, this robustness check categorizes third-party services from Ghostery's "essential" category as trackers instead of an essential third-party service. Consequently, the proportion of websites with "No Tracker Detected" decreases from 62.7% to 55.8%, and the share of websites not keeping their privacy promises increases from 32.9% to 36.3% compared to the main analysis.

# References

1. Bouhoula, A., Kubicek, K., Zac, A., Cotrini, C., Basin, D.: Automated large-scale analysis of cookie notice compliance. In: 33rd USENIX Security Symposium. USENIX Security 24, USENIX Association, Philadelphia, PA, USA (2024). https://www.usenix.org/system/files/sec23winter-prepub-107-bouhoula.pdf
2. Bui, D., Tang, B., Shin, K.G.: Do opt-outs really opt me out? In: Proceedings of the 2022 ACM SIGSAC Conference on Computer and Communications Security, CCS 2022, pp. 425–439. Association for Computing Machinery, New York, NY, USA (2022). https://doi.org/10.1145/3548606.3560574
3. Bujlow, T., Carela-Español, V., Solé-Pareta, J., Barlet-Ros, P.: A survey on web tracking: mechanisms, implications, and defenses. Proc. IEEE **105**(8), 1476–1510 (2017). https://doi.org/10.1109/JPROC.2016.2637878
4. Conference of German Data Protection Authorities: Orientierungshilfe der Aufsichtsbehörden für Anbieter:innen von Telemedien ab dem 1. Dezember 2021 (OH Telemedien 2021) Version 1.1 [Guidance from the supervisory Authorities for Telemedia Providers From December 1, 2021]. Technical report, Conference of German Data Protection Authorities (2022). https://www.datenschutzkonferenz-online.de/media/oh/20221205_oh_Telemedien_2021_Version_1_1_Vorlage_104_DSK_final.pdf
5. Du, X., Yang, Z., Lin, J., Cao, Y., Yang, M.: Withdrawing is believing? Detecting inconsistencies between withdrawal choices and third-party data collections in mobile apps. In: 2024 IEEE Symposium on Security and Privacy. IEEE Computer Society (2024). https://yinzhicao.org/mowchecker/mowchecker.pdf
6. Englehardt, S., Narayanan, A.: Online tracking: a 1-million-site measurement and analysis. In: Proceedings of the 2016 ACM SIGSAC Conference on Computer and Communications Security, CCS 2016, pp. 1388–1401. Association for Computing Machinery, New York, NY, USA (2016). https://doi.org/10.1145/2976749.2978313
7. European Data Protection Board: Guidelines 2/2023 on technical scope of Art. 5(3) of ePrivacy Directive. Technical report, European Data Protection Board (2023). https://www.edpb.europa.eu/system/files/2023-11/edpb_guidelines_202302_technical_scope_art_53_eprivacydirective_en.pdf
8. European Data Protection Board: Opinion 08/2024 on Valid consent in the context of Consent or Pay Models Implemented by Large Online Platforms. Technical report, European Data Protection Board (2024). https://www.edpb.europa.eu/system/files/2024-04/edpb_opinion_202408_consentorpay_en.pdf
9. Fouad, I., Bielova, N., Legout, A., Sarafijanovic-Djukic, N.: Missed by filter lists: detecting unknown third-party trackers with invisible pixels. Proc. Privacy Enhancing Technol. **2020**(2), 499–518 (2020). https://doi.org/10.2478/popets-2020-0038
10. French Data Protection Authority: Cookies: Solutions Pour Les Outils De Mesure D'Audience [Cookies: solutions for audience measurement tools] (2023). https://www.cnil.fr/fr/cookies-et-autres-traceurs/regles/cookies-solutions-pour-les-outils-de-mesure-daudience. Accessed 12 June 2024
11. French Data Protection Authority: LINCnil/CNIL-Cookies-List (2024). https://github.com/LINCnil/CNIL-Cookies-List. Accessed 12 June 2024
12. Ghostery: Ghostery/Ghostery-Tracker-Analytics-Extension: Open sourcing (2023). https://github.com/ghostery/ghostery-tracker-analytics-extension/releases/tag/v0.7.9. Accessed 12 June 2024
13. Ghostery: How Ghostery categorizes trackers (2024). https://www.ghostery.com/blog/how-ghostery-categorizes-trackers. Accessed 12 June 2024

14. Iordanou, C., Smaragdakis, G., Poese, I., Laoutaris, N.: Tracing cross border web tracking. In: Proceedings of the Internet Measurement Conference 2018, IMC 2018, pp. 329–342. Association for Computing Machinery, New York, NY, USA (2018). https://doi.org/10.1145/3278532.3278561

15. Karaj, A., Macbeth, S., Berson, R., Pujol, J.M.: WhoTracks .Me: shedding light on the opaque world of online tracking (2019). https://doi.org/10.48550/arXiv.1804.08959. Preprint on arXiv

16. Koch, S., Wessels, M., Altpeter, B., Olvermann, M., Johns, M.: Keeping privacy labels honest. Proc. Privacy Enhancing Technol. **2022**(4), 486–506 (2022). https://doi.org/10.56553/popets-2022-0119

17. Kollnig, K., Shuba, A., Van Kleek, M., Binns, R., Shadbolt, N.: Goodbye tracking? Impact of iOS app tracking transparency and privacy labels. In: Proceedings of the 2022 ACM Conference on Fairness, Accountability, and Transparency, FAccT 2022, pp. 508–520. Association for Computing Machinery, New York, NY, USA (2022). https://doi.org/10.1145/3531146.3533116

18. Lukic, K., Miller, K.M., Skiera, B.: The impact of the general data protection regulation (gdpr) on online tracking (2023). https://doi.org/10.2139/ssrn.4399388. Preprint on SSRN

19. Matte, C., Bielova, N., Santos, C.: Do cookie banners respect my choice? Measuring legal compliance of banners from IAB Europe's transparency and consent framework. In: 2020 IEEE Symposium on Security and Privacy, pp. 791–809. IEEE Computer Society, San Francisco, CA, USA (2020). https://doi.org/10.1109/SP40000.2020.00076

20. Mayer, J.R., Mitchell, J.C.: Third-party web tracking: policy and technology. In: 2012 IEEE Symposium on Security and Privacy, pp. 413–427. IEEE Computer Society, San Francisco, CA, USA (2012). https://doi.org/10.1109/SP.2012.47

21. Morel, V., Santos, C., Fredholm, V., Thunberg, A.: Legitimate interest is the new consent - large-scale measurement and legal compliance of IAB Europe TCF paywalls. In: Proceedings of the 22nd Workshop on Privacy in the Electronic Society, WPES 2023, pp. 153–158. Association for Computing Machinery, New York, NY, USA (2023). https://doi.org/10.1145/3603216.3624966

22. Müller-Tribbensee, T., Miller, K.M., Skiera, B.: paying for privacy: pay-or-tracking walls (2024). https://doi.org/10.2139/ssrn.4749217. Preprint on SSRN

23. Peukert, C., Bechtold, S., Batikas, M., Kretschmer, T.: Regulatory spillovers and data governance: evidence from the GDPR. Mark. Sci. **41**(4), 746–768 (2022). https://doi.org/10.1287/mksc.2021.1339

24. Rasaii, A., Gosain, D., Gasser, O.: Thou shalt not reject: analyzing accept-or-pay cookie banners on the web. In: Proceedings of the 2023 ACM on Internet Measurement Conference, IMC 2023, pp. 154–161. Association for Computing Machinery, New York, NY, USA (2023). https://doi.org/10.1145/3618257.3624846

25. Skiera, B., Miller, K., Jin, Y., Kraft, L., Laub, R., Schmitt, J.: The impact of the general data protection regulation (gdpr) on the online advertising market. Self-Published (2022). https://www.gdpr-impact.com/

26. Utz, C., Amft, S., Degeling, M., Holz, T., Fahl, S., Schaub, F.: Privacy rarely considered: exploring considerations in the adoption of third-party services by websites. Proc. Privacy Enhancing Technol. **2023**(1), 5–28 (2023). https://doi.org/10.56553/popets-2023-0002

27. VG Wort: Teilnahmebedingungen für das Online Meldesystem T.O.M. der VG WORT (Stand November 2023) [participation conditions for the online reporting system t.O.M of vg wort (as of november 2023)] (2023). https://tom.vgwort.de/portal/showParticipationCondition. Accessed 12 June 2024

# Data Governance and Neutral Data Intermediation: Legal Properties and Potential Semantic Constraints

Emanuela Podda(✉)[iD]

Department of Computer Science, SPDP Lab, Università degli Studi di Milano, Via Celoria 18, 20133 Milan, Italy
emanuela.podda@unimi.it

**Abstract.** The EU Regulation 2022/868 (Data Governance Act) designs a European data governance framework to facilitate data sharing, shaping the role of the Data Intermediation Service Provider. This paper aims to clarify, from a levelism perspective, the main features of the Data Governance Act framing it as a macro-level model, and to explore the data intermediation service provider's role as one of multiple subjective stand-points in this data governance model. In doing so, it attempts to capture the main legal properties of data intermediation service and it brings them to the fore, identifying potential semantic constraints that may need to be addressed when engineering a data governance model in line with legally desirable outcomes. This paper builds on the existing cross-sectoral literature at the intersection of law and technology by relying on the research method of legal analysis and lays the groundwork for implementing a data governance model that integrates and consolidates different levels, dimensions, and facets of the in-force legal framework.

**Keywords:** Data Governance Act · Data Intermediation · Neutral Intermediation

## 1 Introduction and Context

Cross-sectoral research aims at reconciling the different perspectives that are typical of the discourse community in which they operate. Both law and technology feature different types of contributions. However, modern societies pose challenges that present many facets whose investigation requires a joint effort from different communities.

Rooted in between IT decisions domain and management, data governance - and its design - is considered a modern challenge [9] touching upon different

This work was supported by project SERICS (PE00000014) under the MUR NRRP funded by the EU - NextGenerationEU. Views and opinions expressed are however those of the authors only and do not necessarily reflect those of the European Union or the Italian MUR. Neither the European Union nor Italian MUR can be held responsible for them.

M. Jensen et al. (Eds.): APF 2024, LNCS 14831, pp. 189–202, 2024.
https://doi.org/10.1007/978-3-031-68024-3_10

scientific domains. More than ten years ago, it has been framed as a *serious* [21] need of organizations to derive business value from data assets. Although, from a legal perspective, caution is needed when considering data as an asset whose control is asserted through ownership [30], various scientific communities have produced extensive literature on data governance in the last decades.

Scholars [1,2,12,25,35] proposed several methodologies and models to conceptualize data governance from different scientific perspectives, but terminologies and taxonomies to foster its design are still in flux.

In September 2023, the European Regulation 868/2022 - also known as Data Governance Act (DGA) - became applicable to facilitate data sharing in the new forthcoming Common European Data Spaces ecosystem. This Regulation introduces a macro-level data governance relying on the role of data intermediation service providers to which the legislator mandates several requirements, among which *neutrality*, on the implicit assumption that this requirement increases trust in data sharing.

This paper aims to render explicit the role of data intermediaries and the service they provide, as well as the neutrality requirement, clearing the ground from legal uncertainty. To this purpose, it builds on the existing cross-sectoral literature at the intersection of law and technology and, with the research method of legal analysis, clarifies the main features of data intermediation services. Borrowing concepts from IT and legal domains, this paper interconnects and integrates the two perspectives on data governance stemming from these two scientific domains. The aim is to build on the existing conceptual foundations that will help to bring about new insights on the data governance model designed in the DGA, here framed as a macro-level overarching model.

In pursuing this aim, this paper is inspired by the levelism approach used in IT literature by Davidson [9] and Bodò [5] recognizing that neither exclusively top-down nor exclusively bottom-up approaches are likely to be effective in data governance. Their research lines seem to mirror what is considered and already acknowledged by scholars from a legal and regulatory perspective on digital governance. Pagallo [26] highlights that, beyond top-down and bottom-up approaches to digital governance, the middle-out approach emerges as a new model for data protection, artificial intelligence, and the web of data, where forms of co-regulation participate in shaping the governance.

In this regard, in fact, both Davidson [9] and Bodò [5] recognize that data governance is composed of different planned and emergent components, which in turn are composed of other interactions that take place at different levels: at macro-level, at meso-level and at micro-level. Davidson [9] acknowledges that data governance is *nested across societal levels* and that research has primarily focused on the meso-level (organisational/corporate) and micro-level (individual). Hence, Davidson calls for new research lines that encompass the macro-level and go beyond organisational and corporate boundaries. Bodò [5] frames (personal) data governance referring to three different horizontal levels ranging from macro-level strategic frameworks (set by globally competing countries), to meso-level (adopted by data-sharing organizations such as companies and municipalities), to individual strategies at the micro-level. Bodò highlights that meso-level governance is under the dual pressure of the macro-level and the micro-level. Given this, Bodò considers that data intermediaries may empower meso-level governance at best calling for a better definition and delineation of properties of the meso-level data governance within the EU context.

Borrowing the levelism from Davidson [9] and Bodò [5], this paper attempts to explore data governance from a levelism perspective, interconnecting cross-disciplinary literature and building on it. Moreover, although Bodò's [5] levelism perspective is mainly on personal data, this paper considers that, sometimes, non-personal data may require the same protection of personal data, such as in the case of confidential and commercially valuable data.

The paper is structured as follows. Section 2, presents a short theoretical background on the most recent perspectives on data governance recalled in legal and IT literature, introducing the DGA from a levelism perspective. Section 3 focuses on data intermediation as the key building block of the European Data Governance, capturing its main legal properties by analyzing active and passive intermediation, as well as the neutrality requirement and its trust assumption. Drawing from the analysis in Sect. 3, Sect. 4 turns to the discussion on the challenges brought by engineering regulatory requirements from the subjective standpoint of data intermediaries. The paper concludes by proposing to further investigate cross-sectoral methods aimed at overcoming semantic constraints posed by legal neutrality and trust assumption in the DGA, hence integrating and consolidating laws, regulations, and business processes relevant for designing a data governance model compliant with legal rules and principles.

## 2    Data Governance as a Cross-Sectoral Research Domain

The normative dimension of governance imposes a shift from a more centralized approach to rule-making, the so-called Command-and-Control (CaC) where state actors control and enforce laws that are backed by sanctions [4]. In the normative dimension, rule-making is rather decentralized [22]. As such, it requires an allocation of roles and responsibilities within *horizontally* organized structures of state and non-state actors that interact fluently [8], requiring the setting of roles and obligations, as well as (legal) liabilities and (ethical) responsibilities. Hence, it recalls forms of participatory and relational interactions aimed at regulating a given system [35].

According to Viljoen [35], the normative dimension focused on the rule-making approach, represents the main legal root of data governance. However, Floridi [19] highlights that, in the last decades, the advent and rapid development of information and communication technologies fostered scholars from many disciplines to theorize the governance of the digital.

Different scientific communities investigated data governance with the aim of drawing specific frameworks (or models) in governing data, demonstrating that data governance goes far beyond corporate organizational boundaries. Producing extensive literature on the topic, undoubtedly, each research community featured a different contribution and seems not to share a unique definition of *data governance*. For the purpose of this paper, legal and IT contributions are both relevant.

On the one hand, within recent legal literature mainly focused on privacy and data protection, Wong [36] clarifies that data governance presents significant challenges and risks stemming from data features that are linked to the process of *datafication* [34]. Therefore, risk plays a central role. Yet in 2017, while the General Data Protection Regulation was in its roll-out phase, Macenaite [23] highlighted that the debate on data governance and risk regulation was still in its infancy among scholars of governance and regulation, who were beginning to explore the links between digital data, ethics, and risk. Macenaite noticed that, at that time, scholars started *injecting* the risk-based approach in data protection, as a typical instrument of corporate governance, provoking several implications, and future challenges in data protection. These considerations nurtured the policy debate and shaped the lines of accountability in the General Data Protection Regulation, where controllers' duties aim at bench-marking good governance in personal data protection and regulatory compliance. Hence, from a legal perspective, as clarified also by Viljoen [35], data governance commits to developing legal responses to the harm and risks posed to data subjects by datafication, improving the relation between data subjects and data collectors (data-holders).

On the other hand, within recent IT literature mainly focused on information systems, Vial [32,33] highlights that data governance, as part of the digital transformation, aims at achieving a dual scope: creating value and protecting data. His work stresses the fact that data management techniques are not sufficient anymore in modern digital ecosystems and new emerging scenarios, as data are no longer an output of a certain process enhanced by technological artifacts. Within the same domain, Otto [24] defines data governance as *a possible approach* to meet business challenges and requirements among which compliance with regulatory and legal provisions, specifying *who makes decisions concerning data, and the tasks and duties resulting from such decisions.*

Therefore, in attempting to interconnect these two perspectives, it may be reasonable to consider that data governance was born as a *business affair* in the corporate domain that further encompassed a legal-tech dimension evolved into security, data protection, and privacy [29,31]. Such development, as well as its contextualization in modern data-driven societies [28], seems to confirm that, as already highlighted by Bennett [3], data governance deploys its effects in many different domain dimensions, but also within many different societal levels.

Perhaps as a backlash of the subject evolution and its growing importance, the European legislator integrates data governance within its legislative action.

## 2.1 The European Regulation 868/2022: A Macro-level Data Governance Model

In 2020, the European Commission released the European Strategy for Data [15] road-mapping a new legislative wave aimed at implementing ten *Common Euro-*

*pean Data Spaces* for granting availability of large pools of data in strategic economic sectors and domains of public interest. Regulatory implementation imposes to address data processing operations in these new ecosystems where roles and responsibilities should be clear and where new technical solutions and tools for pooling, accessing, and sharing data must ensure legal and regulatory compliance.

Within this strategic setting, the Data Governance Act (EU 868/2022, DGA) is conceived as a horizontal legislation complementing the sectoral one. It introduces: (a) conditions for the re-use of data held by public bodies, (b) role and responsibility of data intermediation services, (c) role and responsibility of entities collecting and processing data for altruistic purposes, (d) role and responsibility of the European Data Innovation Board.

From a levelism perspective, this legislative body defines a macro-level data governance. In this respect, two further considerations are needed. First, the DGA encourages the development of harmonised approaches and harmonised processes in data sharing, even though Member States may have in place different meso-level and micro-level data governance models. In this regard, Davidson [9] emphasises the need to explore various data governance and stakeholder values with cross-disciplinary research. Second, as already highlighted by Bodò [5], even before macro-level data governance emerged, various stakeholders had already defined their approaches on how to govern their data within their business, thus reflecting their business interests and their technical and computational capacity.

In fact, the Regulation Impact Assessment [16] recalls low trust in data sharing as one of the key drivers for the enactment of the Regulation. Specifically, it referred to the fact that the lack of trust increased transaction costs, rendering it difficult to find suitable data exchange partners and to develop interoperability solutions for cleaning, transforming, and sharing data.

Therefore, given this challenging scenario, the DGA was conceived as a new macro-level model of data governance, not opposed but, *juxtaposed* to the industry-based (meso-level) model developed by large market players holding a consolidated market power and large amounts of data. In this new macro-level model, the main building block is the envisaged role of data intermediation service providers as intermediaries who *mediate*, through commercial relationships, the interactions between data-holders and users in a neutral fashion way. According to the Regulation, this mediation takes place within a legally typified possible action: *facilitating data sharing* and *undertaking data sharing*.

## 3  Data Intermediation-as-a-Service: A Subjective Approach to Data Governance

The DGA defines the data intermediation service in Article 2(11) as *a service which aims to establish commercial relationships for data sharing between an undetermined number of data subjects and data-holders on the one hand and data users on the other, through technical, legal or other means [...].*

Yet, a literal interpretation of the article makes it possible to capture two main distinguishing properties of the service: the scope and the features of subjects involved in the mediation service.

The scope corresponds to the establishment of commercial relationships for sharing data. In this regard, the Regulation does not define the intended meaning of commercial relationships but, as already highlighted by Carovano [7], it should be intended as to where the service is provided against payment. As a consequence, this requirement determines what services cannot fall under the Regulation. For instance, as also generally recalled in Recital 28 (supporting interpretation), providers of cloud storage and data sharing software offering the service for free, cannot be considered offering intermediation services *ex* DGA.

Concerning the features of subjects involved in the intermediation service, the Regulation refers to *an undetermined number* of data-holders/subjects and data users. In this regard, Recital 28 clarifies that providers of cloud services, data-sharing software, web browsers, or email services are excluded from the Regulation scope of the application. Recital 27 in conjunction with Article 10 further clarifies that the intermediation service may include bilateral, as well as multilateral, sharing of data or the creation of platforms or databases enabling the sharing or joint use of data, as well as the establishment of specific infrastructure aimed at interconnecting data subjects and data-holders with data users. In line with these clarifications, Carovano [7] considers that the legal reference to *an undetermined number* seems not to be intended for the design and deployment of a user metric in providing the intermediation service. Rather, from his point of view, it seems to refer to the case in which the intermediation service corresponds to the implementation of conditions of access to the service.

Moreover, it is relevant to consider that the Regulation maintains a technologically agnostic approach as it does not favor any specific technology for the provision of the service: according to the letter of the law, the type of intermediation service may vary. In this specific regard, Article 2 lists the intermediation means as *technical, legal, or other means [...]*. In addition, the letter of the law recalls intermediation activities, but also to activities aimed at *facilitating intermediation*. This lexicon seems to provide legal endorsement in the design of intermediation services that may include, to name a few, access control policies, data model transfers (i.e. federated learning), legal agreements on data transactions, and license agreements.

In this regard, the European Union Agency for Cybersecurity (Enisa) [18], highlights that data intermediaries *mediate* the trustful sharing of data, ensuring a secured and controlled sharing. This being given, the intermediary roles and responsibilities seem to depend on the processing operations, on how the sharing takes place, and on the kinds of risk vectors that may arise from the sharing and processing operations. In this respect, there is no one-size-fits-all solution, and clarifying how the data flows between these entities is key for designing a data governance model that ensures legal compliance.

## 3.1   Active and Passive Intermediation

Article 10 defines data intermediation services as:

*(a) intermediation services between data-holders and potential data users, including making available the technical or other means to enable such services; those services may include bilateral or multilateral exchanges of data or the creation of platforms or databases enabling the exchange or joint use of data, as well as the establishment of other specific infrastructure for the interconnection of data-holders with data users;*

*(b) intermediation services between data subjects that seek to make their personal data available or natural persons that seek to make non-personal data available, and potential data users, including making available the technical or other means to enable such services, and in particular enabling the exercise of the data subjects' rights provided in Regulation (EU) 2016/679;*

*(c) services of data cooperatives.*

Enisa [17] has already highlighted that data-holders may assign to intermediaries different roles and responsibilities, according to the different privacy and security goals that data-holders want to achieve, some of which may be more active than others.

In line with this consideration, it seems reasonable to distinguish between active and passive intermediation (Table 1), depending on whether the intermediary handles raw-level data-holder data. This would imply that, within the wide range of intermediation activities, passive intermediation would be limited to those activities where the intermediary does not handle the data-holder data, i.e. is not actively involved in data transactions, but rather has access to metadata only. On the contrary, active intermediation would imply that the intermediary handles raw-level data-holder data, mediating by transacting data-holder data. While the former (passive intermediation) is loosely defined in the Regulation as *facilitating data sharing*, the latter (active intermediation) is defined as *undertaking data sharing*.

**Table 1.** Data Intermediary Service *ex* Data Governance Act

| Type of Intermediation | |
| --- | --- |
| Active | Passive |
| Access to Metadata | Access to Metadata |
| Access to Raw-Level Data-Holder Data | No Access to Raw-Level Data-Holder Data |

Active intermediation may be envisaged when the intermediary transacts data between the parties *undertaking data sharing*, handling data-holder data, and undertaking processing activities as well on raw-level data-holder data. In this case, the Regulation specifies that these services go beyond the mere facilitation of data sharing, provided that they are explicitly requested or approved by data-holders. According to the letter of the law, in Art. 12.1(e), this may be the case when the intermediary offers additional services such as temporary storage, curation, conversion, anonymization, or pseudonymization of data.

Differently, passive intermediation may be envisaged when the intermediary does not have access to raw-level data-holder data but the intermediation activities aim at *facilitating data sharing*. This may be the case when, as generally recalled in Art. 12(c) of the Regulation, data intermediation service may entail the use of data for the detection of fraud or cybersecurity. Relying on a legal agreement, the intermediary service provider may be required to perform fraud prevention and detection, providing a first layer of security by, for instance, monitoring the sources of access requests, analyzing the log activity performed by the data user, and transmitting only the trusted request to the data-holder. Another example of passive intermediation may be the case in which the intermediary service provider facilitates the intermediation exclusively by legal means, such as supporting the parties in defining terms and conditions of the Service Level Agreements, or the Data Sharing Agreements or, most generally in drafting commercial licenses. To confirm the feasibility of this case, it should be recalled that the Data Governance Act aims at creating a *juxtaposed* data governance model to promote data sharing and data access for European SMEs, whose computational power and technical capacity cannot be compared to those of the big tech market players.

Moreover, even if not explicitly mentioned by the Regulation, it is relevant - for the aim of this paper - the case in which the computation would take place in trusted and secure environments. Here, data are kept in the data-holder's premises and users would be allowed to run their computations over raw-level data-holder data, with restricted access. In this case, the intermediary exerts strict monitoring and control on the accesses by preventing data users from importing or exporting data. In such a case, the intermediary will be the one setting the trusted environment, due to the reduced computational capacity of the data-holder. Security reasons may justify relying on a different infrastructure from the one where the data-holder's data is usually hosted.

In all these cases, compliance with the confidentiality duty by the intermediary service should be a must. On the one hand, its role is to smooth out information and power asymmetries as well as bargaining power between the parties involved in data transactions and computations. On the other hand, the development of technical solutions to minimize the risk posed to data confidentiality should be considered a top priority, not only from a General Data Protection perspective. The intermediary may handle not only personal data but also other data whose confidentiality should be protected on other legal grounds.

## 3.2 Neutrality Requirement and Trust Assumption

Article 12 mandates intermediation service providers to respect fifteen conditions - substantial and procedural - when providing their service. Among these, one of the substantial requirements is loosely mentioned in Recital 33 as *neutrality*. The Recital clarifies that neutrality is a key element in enhancing trust when intermediating interactions between data-holders and data users. Article 12.1(a) further determines its extent, imposing that *[...] the data intermediation services provider shall not use the data for which it provides data intermediation services for purposes other than to put them at the disposal of data users and shall provide data intermediation services through a separate legal person; [...]*.

Literal interpretation suggests that neutrality impacts trust and risk vectors when the intermediary provides the service. However, further reflection is needed to assess the extent to which this legal requirement alone allows to achieve the highest level of trust in the intermediary. This is mainly because, from a technical standpoint, the level of trust the data-holder places over the intermediary service provider determines the type of computations that the intermediary service provider is legitimated to perform over data-holder data.

From a legal standpoint, providing data intermediation services through a separate legal entity may have a positive impact on the allocation of roles by facilitating the identification of responsibilities and liabilities among the parties involved in the intermediation. In addition, the provision of the service through a separate legal entity poses additional questions as to whether its nature should be public or private. In this regard, the Regulation does not impose any specific legal nature.

The question is whether legal neutrality, *per se*, may be sufficient in increasing trust among parties. The Data Governance Act is intended to bind interactions and processing operations relying on high-performance computing and cloud federation. As stressed by Institutional documents [15], cloud federation allows for promoting a *gradual rebalancing between centralized data infrastructure in the cloud and highly distributed and smart data processing at the edge [...]*.

In fact, within a cloud federation scenario, processing activities and computations may be collaborative and distributed implying different impacts on data security and privacy; addressing risk vectors may be challenging. In cloud scenarios, risk variations depend on the architectural or trust assumption: different deployment models can range from private cloud, public cloud, community cloud, and hybrid cloud. Trust assumptions shape the level of confidentiality and the extent of the confidentiality duty binding the service provider when providing the service. Regardless of the different deployment models, IT scholars [10] have already stressed the need to protect data from the cloud service provider itself: providers should not know the actual content of data.

Within the context envisaged by the DGA, an intermediary service provider could either be a cloud service provider or, if not, may rely on cloud services for intermediation activities. In both cases, data confidentiality should be guaranteed as a main legally desirable outcome whose compliance assessment may vary according to different laws and regulations in force. Recently, De Capitani di Vimercati [11] investigates the main issues raised in cloud-based storage and processing, highlighting security and privacy risks and challenges posed by outsourcing data and computations to cloud services.

In light of these considerations, it is reasonable to consider that legal neutrality may pose semantic constraints in the design of a data governance model. In fact, despite the European legislator conceives legal neutrality as high trust assurance on the intermediary service provider, IT scholars highlight the risks posed to privacy and security in cloud-based scenarios.

Addressing these potential semantic constraints may require the integration of additional regulatory frameworks binding the Intermediary Service Provider in its intermediation activity. In this regard, further analysis is required on the interplay between the DGA and additional regulatory frameworks that may be related to the semantics of the intermediated data, thus whether personal or non-personal, as well as related to the liability and due diligence for data processing or data storage in cloud-based services.

## 4   The Interplay of Law and Technology in Intermediation Service *ex* DGA

The previous sections proposed a legal analysis to assess and capture the main legal properties of the data intermediation service as envisaged in the DGA. Based on semantic interpretation, the legal analysis interconnected law and technology, to further facilitate the design of a data governance model by embedding legal constraints and safeguards [27] in the aims at legal compliance.

Over the past few decades, researchers have produced substantial literature on modeling legal rules especially on the automation of personal data protection principles to ensure compliance while detecting violations of the GDPR. Recently, scholars [14] proposed ontological engineering and semantic technologies to implement the DGA; this can also support the design of a data governance model.

However, as highlighted in the legal analysis and stressed in the research lines from which this paper draws inspiration [5,9], data governance takes place at multiple levels challenging the caption of legal requirements and its correlated technical implementation. Among IT scholars working at the intersection of technology and regulatory compliance, Boella [6] highlights that legal knowledge cannot be considered a *monolithic entity* because legal elements may change when considering different legal levels and domains.

In this specific regard, the analysis of the macro-level data governance model and the subjective standpoint (data intermediary service provider) confirms that legal requirements depend on many dimensions and facets of the legal system. The legal system is inherently multilevel, multidimensional, and multifaceted. Variations of regulatory dimensions and facets may depend not only on the subjective standpoints of the stakeholders bound by a specific regulatory framework but also on objective technical contexts regulated by a different framework. It seems not superfluous to recall here what has already been highlighted by law scholars, namely Hildebrandt [20] when considering that the nature of legal language may be perceived by non-legal experts as misleading, as a bug in the system.

To this purpose, as also recalled by the Regulation, investigating legal interoperability among different levels, contexts, and facets seems to be key. Yet a decade ago, Boella [6] considered legal semantic interoperability as *one of the main issues about knowledge engineering*. This implies that even in a generally applicable overarching macro-level governance, roles, duties, responsibilities, and liabilities in a given system are not perfectly crystallized. These features and properties may stem from diverse legal and regulatory frameworks deploying their effect at different levels. Their integration should represent the first step to determining the legally desirable outcomes, often resulting from a balancing exercise of competing interests [13].

In line with these considerations, the combination and integration of multi-leveled, multi-layered, and multi-faceted laws and regulations is a necessary step for designing a data governance model that embeds legal norms and/or policies. This paper explored the DGA as the European macro-level model of data governance. By clarifying the role of the data intermediary as a key socio-technical component in the considered framework, it attempted to set the ground for the design of a data governance model that embeds legal norms and policies derived from the DGA and highlights its limitations.

## 5   Conclusion and Further Research Directions

Models are conceived as abstract representations of a target system and capture a part of reality depending on the level of abstraction used in the description of the targeted system. Typically, the analysis of the model features is left implicit in the modeling process.

Intending to render explicit these features, the paper describes and explores the intermediation service to enrich the ontological understanding of intermediation activity *ex* DGA. The legal analysis highlights that this regulatory framework presents specific semantic constraints in the case of legal neutrality and trust assumptions. Given this limitation, the author considers that this potential semantic constraint could be addressed by integrating additional regulatory frameworks binding the Intermediary Service Provider in its intermediation activity. In this regard, further analysis is required on the interplay between the DGA and other additional and relevant regulatory frameworks.

Further research integrating multiple levels of abstraction is needed to better capture the features of the legal systems and derive a data governance model that grants privacy-by-design, as well as security-by-design in the intermediated data transactions.

## References

1. Abraham, R., Schneider, J., Vom Brocke, J.: Data governance: a conceptual framework, structured review, and research agenda. Int. J. Inf. Manage. **49**, 424–438 (2019). https://doi.org/10.1016/j.ijinfomgt.2019.07.008

2. Benfeldt, O., Persson, J.S., Madsen, S.: Data governance as a collective action problem. Inf. Syst. Front. **22**, 299–313 (2019). https://doi.org/10.1007/s10796-019-09923-z

3. Bennett, C.J., Raab, C.D.: Revisiting the governance of privacy: contemporary policy instruments in global perspective. Regul. Gov. **14**(3), 447–464 (2020). https://doi.org/10.1111/rego.12222. _eprint: https://onlinelibrary.wiley.com/doi/pdf/10.1111/rego.12222

4. Black, J.: Decentring regulation: understanding the role of regulation and self-regulation in a 'post-regulatory' world. Curr. Leg. Probl. **54**(1), 103–146 (2001). https://doi.org/10.1093/clp/54.1.103

5. Bodó, B., Irion, K., Janssen, H., Giannopoulou, A.: Personal data ordering in context: the interaction of Meso-level data governance regimes with macro frameworks. Internet Policy Rev. **10**(3) (2021). https://doi.org/10.14763/2021.3.1581

6. Boella, G., Rossi, P.: The multi-layered legal information perspective. In: Sartor, G., Casanovas, P., Biasiotti, M., Fernández-Barrera, M. (eds.) Approaches to Legal Ontologies: Theories, Domains, Methodologies, vol. 1, pp. 133–141. Springer, Dordrecht (2011). https://doi.org/10.1007/978-94-007-0120-5_8

7. Carovano, G., Finck, M.: Regulating data intermediaries: the impact of the data governance act on the EU's data economy. Comput. Law Secur. Rev. **50**, 105830 (2023). https://doi.org/10.1016/j.clsr.2023.105830

8. Colebatch, H.: Making sense of governance. Policy Soc. **33**(4), 307–316 (2014) https://doi.org/10.1016/j.polsoc.2014.10.001. _eprint: https://doi.org/10.1016/j.polsoc.2014.10.001

9. Davidson, E., Wessel, L., Winter, J.S., Winter, S.: Future directions for scholarship on data governance, digital innovation, and grand challenges. Inf. Organ. **33**(1), 100454 (2023). https://doi.org/10.1016/j.infoandorg.2023.100454

10. De Capitani di Vimercati, S., Erbacher, R.F., Foresti, S., Jajodia, S., Livraga, G., Samarati, P.: Encryption and fragmentation for data confidentiality in the cloud. In: Aldini, A., Lopez, J., Martinelli, F. (eds.) FOSAD 2012-2013. LNCS, vol. 8604, pp. 212–243. Springer, Cham (2014). https://doi.org/10.1007/978-3-319-10082-1_8

11. De Capitani Di Vimercati, S., Foresti, S., Samarati, P.: Protecting data and queries in cloud-based scenarios. SN Comput. Sci. **4**(5), 440 (2023). https://doi.org/10.1007/s42979-023-01862-6

12. Delacroix, S., Lawrence, N.D.: Bottom-up data trusts: disturbing the 'one size fits all' approach to data governance. Int. Data Privacy Law **9**(4), 236–252 (2019). https://doi.org/10.1093/idpl/ipz014

13. Durante, M.: Dealing with legal conflicts in the information society. An informational understanding of balancing competing interests. Philosophy Technol. **26**(4), 437–457 (2013). https://doi.org/10.1007/s13347-013-0105-z

14. Esteves, B., Victor, R.D., Pandit, H.J., Lewis, D.: Semantics for implementing data reuse and altruism under EU's data governance act. In: Knowledge Graphs: Semantics, Machine Learning, and Languages, pp. 210–226. IOS Press (2023). https://doi.org/10.3233/SSW230015

15. European Commission: Communication from the Commission to the European Parliament, the Council, the European Economic and Social Committee and the Committee of the Regions - A European strategy for data COM/2020/66 final (2020). https://eur-lex.europa.eu/legal-content/EN/TXT/?uri=CELEX%3A52020DC0066

16. European Commission: Impact Assessment Report Accompanying the document Proposal for a Regulation of the European Parliament and of the Council on European Data Governance (Data Governance Act) COM(2020) 767 final - SEC(2020) 405 final - SWD(2020) 296 final, November 2020. https://eur-lex.europa.eu/legal-content/HR/TXT/?uri=CELEX:52020SC0295

17. European Union Agency for Cybersecurity: Engineering personal data protection in EU data spaces. Technical report, Publications Office, LU (2024). https://doi.org/10.2824/210862

18. European Union Agency for Cybersecurity, ENISA: Engineering Personal Data Protection in EU Data Spaces. Report/Study, January 2024. https://www.enisa.europa.eu/publications/engineering-personal-data-protection-in-eu-data-spaces

19. Floridi, L.: Soft ethics and the governance of the digital. Philosophy Technol. **31**(1), 1–8 (2018). https://doi.org/10.1007/s13347-018-0303-9

20. Hildebrandt, M.: Grounding computational 'law' in legal education and professional legal training. In: Research Handbook on Law and Technology, pp. 99–127. Edward Elgar Publishing, December 2023. https://doi.org/10.4337/9781803921327.00014

21. Khatri, V., Brown, C.V.: Designing data governance. Commun. ACM **53**(1), 148–152 (2010). https://doi.org/10.1145/1629175.1629210

22. Levi-Faur, D.: The Oxford Handbook of Governance. Oxford Universit Press (2012)

23. Macenaite, M.: The "Riskification" of European data protection law through a two-fold shift. Eur. J. Risk Regul. **8**(3), 506–540 (2017). https://doi.org/10.1017/err.2017.40

24. Otto, B.: Data governance. Bus. Inf. Syst. Eng. **3**(4), 241–244 (2011). https://doi.org/10.1007/s12599-011-0162-8. http://link.springer.com/10.1007/s12599-011-0162-8

25. Otto, B.: A morphology of the organization of data governance. In: ECIS 2011 Proceedings, p. 272 (2011). https://aisel.aisnet.org/ecis2011/272

26. Pagallo, U., Casanovas, P., Madelin, R.: The middle-out approach: assessing models of legal governance in data protection, artificial intelligence, and the web of data. Theory Pract. Legislation **7**(1), 1–25 (2019). https://doi.org/10.1080/20508840.2019.1664543

27. Pagallo, U., Durante, M.: The pros and cons of legal automation and its governance. Eur. J. Risk Regulation **7**(2), 323–334 (2016). https://doi.org/10.1017/S1867299X00005742

28. Pentland, A.: The data-driven society. Sci. Am. **309**(4), 78–83 (2013). https://www.jstor.org/stable/26018109

29. Power, E.M., Trope, R.L.: The 2006 survey of legal developments in data management, privacy, and information security: the continuing evolution of data governance. Bus. Lawyer **62**(1), 251–294 (2006). https://www.jstor.org/stable/40688420

30. Rule, J., Hunter, L.: 8. towards property rights in personal data. In: 8. Towards Property Rights in Personal Data, pp. 168–181. University of Toronto Press, March 1999. https://doi.org/10.3138/9781442683105-010

31. Trope, R.L., Power, E.M.: Lessons in data governance: a survey of legal developments in data management, privacy and security. Bus. Lawyer **61**(1), 471–516 (2005)

32. Vial, G.: Understanding digital transformation: a review and a research agenda. J. Strateg. Inf. Syst. **28**(2), 118–144 (2019). https://doi.org/10.1016/j.jsis.2019.01.003

33. Vial, G.: Data governance and digital innovation: a translational account of practitioner issues for IS research. Inf. Organ. **33**, 100450 (2023). https://doi.org/10.1016/j.infoandorg.2023.100450

34. Mayer-Schonberger, V., Cukier, K.: Big Data: A Revolution That Will Transform How We Live, Work and Think. Houghton Mifflin Harcourt, Canada (2013)

35. Viljoen, S.: A relational theory of data governance feature. Yale Law J. **131**(2), 573–654 (2021)

36. Wong, W.H., Duncan, J., Lake, D.A.: Why data about people are so hard to govern. Regul. Gov. **n/a**(n/a) (2024). https://doi.org/10.1111/rego.12591. _eprint: https://onlinelibrary.wiley.com/doi/pdf/10.1111/rego.12591

# No Transparency for Smart Toys

Julika Feldbusch⬤, Valentyna Pavliv⬤, Nima Akbari⬤,
and Isabel Wagner$^{(\boxtimes)}$⬤

University of Basel, Spiegelgasse 1, 4054 Basel, Switzerland
julika.feldbusch@stud.unibas.ch,
{valentyna.pavliv,nima.akbari,isabel.wagner}@unibas.ch
https://pet.dmi.unibas.ch/en/

**Abstract.** Smart toys combine traditional playtime with modern technologies, integrating IoT features like communication, computation, and sensing to create interactive toys that respond to their environment, offering children new options for entertainment and playful education. However, despite well-documented privacy and security shortcomings of IoT devices, there are no recent studies on the privacy and security properties of smart toys. This is critical because children are a particularly vulnerable group whose personal data merits special protection. In this paper, we therefore examine 12 smart toys available in the EU market with regard to their security, privacy, and transparency. Our main findings include widespread behavioral profiling of children via toy analytics data and a lack of transparency due to insufficient and not easily accessible information about data collection and processing.

**Keywords:** Smart toys · Transparency · Privacy · Security · GDPR

## 1 Introduction

In 2015, the Hello Barbie doll was released as the first toy that children could speak with in a meaningful way, thanks to a direct connection to a remote server running speech processing software [32]. The Hello Barbie servers stored each doll's conversation history, which enabled the doll to build conversations based on previous interactions. Hello Barbie was one of the first in a generation of *smart toys*: toys with communication, computation, and sensing capabilities similar to other Internet of Things (IoT) devices. Concerns by parents were followed by a series of academic studies that detailed the security and privacy issues with Hello Barbie [7,24,34]. Subsequently, Mattel discontinued Hello Barbie in 2017.

Since the scandals about Hello Barbie and a similar doll, My Friend Cayla, we are not aware of major public scandals involving other smart toys. One reason for this could be the adoption of the General Data Protection Regulation (GDPR) [17] in 2016, which set strict rules for data protection and privacy. However, questions remain about how rigorously these regulations are applied to children's toys.

Privacy and data protection are especially important for products aimed at children because children's play time is essential for development as a human being [4,58], and pervasive surveillance by cameras and microphones in toys is

M. Jensen et al. (Eds.): APF 2024, LNCS 14831, pp. 203–227, 2024.
https://doi.org/10.1007/978-3-031-68024-3_11

detrimental to this process. Importantly, the degree of surveillance that a child is exposed to should not depend on well-informed and privacy-conscious parents who are able to judge the implications of pervasive surveillance.

Although numerous smart toys are available on the market today, we are not aware of recent studies that investigate their privacy or security. In this paper, we close this gap by rigorously examining 12 smart toys, aiming to understand to what extent the security, privacy, and transparency properties of currently available toys have improved, following the European Telecommunications Standards Institute (ETSI) *Cybersecurity for Consumer IoT* [16] standard as guideline. We purchase a diverse selection of 12 toys and evaluate their *security properties* by examining encrypted and unencrypted traffic, cipher suites and TLS versions, as well as Wi-Fi security; *privacy properties* by analyzing the transmitted data and requested application permissions, by decrypting the traffic or inspecting the application's code; *transparency properties* by inspecting privacy policies and sending subject access requests as well as data deletion requests. We also briefly comment on the toys' likely *compliance* with the GDPR and the upcoming Cyber Resilience Act (CRA). Our main findings are:

1. *Security* is addressed well overall, even though most toys still use TLS 1.2 encryption and only some use forward-secrecy cipher suites. In local networks no data traffic is encrypted, resulting into an insecure Wi-Fi setup for toy devices without user interface. None of our toys specify a hardware/software support period, and some toys even state that they may discontinue service without notice.
2. *Privacy* protections are generally insufficient. Most toys collect extensive analytics data combined with unique identifiers, subjecting children to pervasive behavioral profiling. In addition, the required permissions of companion apps are often unnecessary and sensitive, such as access to location, contacts, or the microphone.
3. *Transparency* suffers from well-known usability issues of privacy policies. The process for sending a subject access request is often too complicated. In addition, only 43% of toys vendors respond to our subject access request within the one month period allowed by GDPR, with one providing incomplete information. None of the toys provide accessible information about security, security issues, or software updates in their user interface.
4. *Compliance* with the GDPR is only partial. At the very least, the availability and comprehensiveness of privacy policies are insufficient. Vendors are also not yet meeting the requirements of the recently approved CRA and may have to improve their processes before the act takes effect.

Based on these findings, we strongly recommend that toy makers prioritize privacy and security, following best practices in security and privacy engineering, to act responsibly towards their young target audience. Subject access requests and data deletion requests should be easier for users and their processing should ideally be automated. Consent should be acquired on an opt-in basis, instead of the default opt-out. It should also be more fine-grained, allowing users to disagree selectively with parts of privacy policy.

The research community and regulators could also do more to support toy makers, for example by standardizing privacy/transparency labels for toy packages, similar to well-known nutrition labels [11,12,19,26,46].

## 2  Related Work

*IoT Security.* Security approaches for IoT devices have to balance security, performance, and resource constraints. Achieving the right trade-off between these factors is a significant challenge, especially given that traditional security practices were optimized for desktop and server environments [35,38].

The two most pressing concerns in IoT security are authentication and encryption. These aspects are essential for safeguarding sensitive data and ensuring secure communications within the IoT ecosystem [49]. Today, most traffic from IoT devices is encrypted, however, data is frequently sent to third-party services and to different geographic regions, which leads to information exposure [47]. Although the connection may be encrypted with the Transport Layer Security (TLS) protocol, the implementation needs to be secure, using strong cipher suites, certificates and protocols. Still, many consumer IoT products lack a secure implementation of TLS, for example by using outdated cipher suites, offering attack surfaces on the connection [43]. Companion apps are often used to control IoT devices or to provide internet connectivity via Wi-Fi or Bluetooth. These apps may also be a security risk, as they may expose user data without proper disclosure [36], use abandoned domains or hard-coded credentials [50]. To determine which data is sent, the TLS encryption of the companion apps to the services can be decrypted using Man-in-the-Middle (MitM) attacks [54] as many vendors have insufficient countermeasures in place [41].

*Transparency.* Current literature on transparency mainly addresses the lack of transparency in privacy requirements [6,13,59], but does not extend to broader cybersecurity metrics. Some studies have proposed the implementation of transparency labels on devices to provide users with a quick overview of basic privacy status [12,19,26,46]. Extending this concept, the idea of labeling could include basic cybersecurity parameters. However, it is important that such labels strike a balance between providing sufficient information for informed decision-making and avoiding overwhelming consumers with excessive detail [39]. Therefore, transparency labels should be designed to be easily understood by non-technical users, while still providing meaningful information.

*Children's Rights.* Children are a particularly vulnerable group that deserves special protection [55]. Especially in the context of online privacy, safeguards for children are being addressed by current legislation. For example, the GDPR in the European Union (EU) provides privacy protections with a focus on parental consent [31]. Recital 38 of the GDPR specifically states that "Children merit specific protection with regard to their personal data, as they may be less aware of the risks, consequences and safeguards concerned" [17].

*Smart Toys Security and Privacy.* Privacy and security of smart toys have been analyzed through both legislative and experimental approaches. Legal studies have found that privacy and security legislation in the EU as well as in North America [22,24,52] does not sufficiently ensure IT security and is difficult to enforce, as many vendors interpret the applicable laws as vaguely as possible. For users of toys, this often means that the toy manufacturers' privacy policies lack clarity or omit information regarding how their toys protect users' security and privacy [7].

Experimental studies have uncovered vulnerabilities primarily caused by a lack of encryption and authentication of Hypertext Transfer Protocol (HTTP) sessions, constant data storage and the disclosure of personal data to third-parties in crash reports [8]. These vulnerabilities lead to privacy policy violations and easily exploitable toys which expose children to various threats, whether through physical, nearby, or remote access to the toy [51]. The identified risks include manipulating a toy's movements or issuing instructions to a child, leading them into dangerous situations. Toys that are disclosing personal information could be turned into spies if exploited as passive sensitive data collectors.

These studies highlight significant shortcomings in cybersecurity and privacy of smart toys. However, it is worth noting that these studies focus mainly on toys which are no longer on the market. Notable examples include the Hello Barbie doll, which was discontinued due to numerous vulnerabilities leading to the potential disclosure of personal data [5,32,51], and the My Friend Cayla doll, which was banned by German authorities due to concerns about a hidden camera and microphone [3,5].

Notably, these experimental studies have been published more than five years ago, with very little work in recent years. However, this most likely does not indicate that the privacy and security issues with smart toys have all been resolved. For example, a recent blog post revealed severe security issues on an educational robot, without disclosing the robot's brand [20]. The uncovered vulnerabilities included APIs that allow stealing of sensitive information, guessable robot identifiers that enable targeted calls to any robot, and remote code execution during the update process.

In this study, we aim to close this gap in research on smart toys and analyze the security, privacy, and transparency properties of toys available for purchase in late 2023.

## 3   Methodology

We follow a three-step methodology to evaluate smart toys with respect to their privacy and security. First, we select and purchase a number of smart toys (Sect. 3.1). Second, we develop a set of evaluation criteria based on the *ETSI Cybersecurity for Consumer IoT* standard (Sect. 3.2). Finally, we select and implement technical methods that allows to evaluate our criteria (Sect. 3.3).

### 3.1 Smart Toys Market and Selection

We select 12 smart toys from the *Toys* category on amazon.de, opting for this marketplace due to its large product range, position in the EU market, and availability of English language information on all toys. We aim for a diverse selection of toys based on their game mechanisms and *smart* features. Regarding connectivity, we only select toys equipped with Wi-Fi capabilities. Among these, eight toys come with built-in connectivity, while four toys use a companion app to access connectivity. Five of these toys are made by well-known toy companies and six by startups. In terms of location, five of the toy vendors are from Europe, four from the United States, and three from Asia. In Table 1 the purchased toys are presented. None of the toys are having artificial intelligence (AI) features because none were available on the EU market when we purchased the toys in October 2023. However, since then several new AI-powered smart toys have become available [10,23,29,33] which we aim to study in future work.

### 3.2 Criteria

To evaluate the security, privacy and transparency properties of smart toys, we follow the *ETSI Cybersecurity for Consumer IoT* [16] standard as a reference. However, due to space constraints, we only present subset of these criteria which represent the key findings of our examination.

*Cybersecurity.* We evaluate the encryption of connections to external servers and on the local network. By capturing the traffic, we investigate whether the connection is encrypted, and if so, what TLS versions and cipher suites are being used. Then we look at the process of initializing a Wi-Fi connection. Since not all toy devices contain user interfaces and open their own local Wi-Fi network, we verify whether encryption is used, and if not, we try to eavesdrop on the connection to retrieve the transmitted data and Wi-Fi credentials. Moreover, we examine the vendor's website, user manual, and terms and conditions to find information about minimum hardware and software support periods.

*Privacy.* To assess the state of children's privacy on the toys, we first evaluate what personal data is sent to which service, by decrypting the data connections from the companion apps to external servers. We then assess the necessity of sending this data, check for the use of known trackers by reviewing the app's codebase and captured network traffic, and examine the destination countries of outgoing traffic. Additionally, we check whether the apps are sending the device's Google Ad ID. In a last step, we obtain the permissions requested by companion apps, by reviewing the app's manifest, and evaluate if the apps request only permissions that are required for their functionality.

*Transparency.* In order to assess adequate user transparency on data collection and use, we send subject data access requests to all vendors on which we have a user account with an e-mail address (7 of the 12 toys). In addition we evaluate the

**Table 1.** Overview of the 12 toys in our study. Popularity of toys estimated by rating numbers on amazon.de (✿: under 500, ✿✿: under 10.000, ✿✿✿: over 10.000)

| Toy | Description | IoT | App | Vendor | Country | Size | Pop. |
|---|---|---|---|---|---|---|---|
| Edurino | Learning app with figurine and pen | - | ✓ | Edurino | DE | | ✿✿ |
| Kidibuzz | Smartphone for children with parental control | ✓ | ✓ | VTech | HK | | ✿ |
| Moorebot | Moving and patroling robot with camera and microphone | ✓ | ✓ | Pilot Laboratories | US | | ✿ |
| Children's Camera | Camera with direct printing | ✓ | ✓ | unknown | CN | ? | ✿✿ |
| Osmo | Learning app with building blocks and camera reflector | - | ✓ | Tangible Play | US | | ✿✿✿ |
| Pictionary Air | Video-capturing drawing app with flashlight stylus | - | ✓ | Mattel | US | | ✿✿✿ |
| Tamagotchi Uni | Virtual pet which can be carried around on wrist | ✓ | - | Bandai Namco | JP | | ✿✿ |
| Tigerbox | Speaker with touchscreen | ✓ | ✓ | Tigermedia | DE | | ✿✿ |
| Tiptoi | Charging station + pen, with audio content for picture books | ✓ | - | Ravensburger | DE | | ✿✿ |
| Toniebox | Speaker with NFC figurine | ✓ | ✓ | Tonies | DE | | ✿✿✿ |
| Twister Air | Video-capturing dancing app with wrist bands | - | ✓ | Hasbro | US | | ✿ |
| Winky | Non-moving robot with gyroscope, program and play | ✓ | ✓ | Mainbot | FR | | ✿ |

availability and comprehensiveness of privacy policies, focusing on information on data collection and transmission. We also evaluate the information presented in user interfaces, regarding software updates, version numbers, and consent and withdrawal options regarding data processing, storage and transmission.

*Compliance.* In the last category, we read the EU legislation regarding data protection (GDPR) and align our results with an assessment on compliance to this piece of legislation. Furthermore, we evaluate if the upcoming Cyber Resilience Act in the current version, would change the current state of smart toys, and preview the potential changes and improvements it could bring.

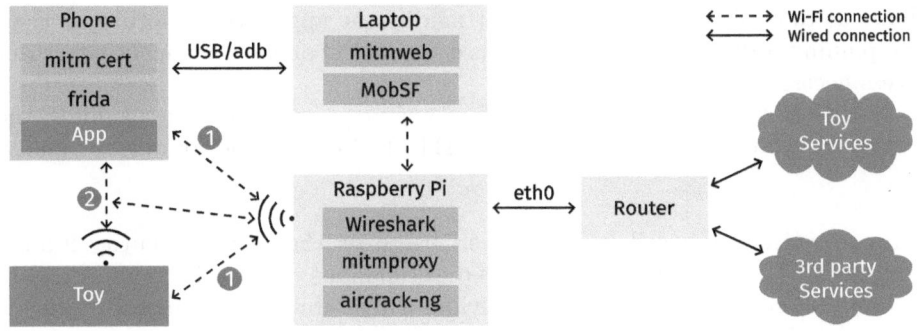

**Fig. 1.** Setup for network traffic analysis. The toy and its companion app connect to the Wi-Fi provided by the Raspberry Pi ❶, which forwards the traffic to the router via an Ethernet connection. Some toys also use a direct Wi-Fi connection to the phone ❷. The Raspberry Pi runs `Wireshark` to capture all traffic, `aircrack-ng` to sniff traffic between the toy and the phone, and `mitmproxy` to decrypt the traffic. The laptop connects to the phone with USB to install the `mitmproxy` certificate and the `Frida` server, and to perform static analysis with `MobSF`.

### 3.3 Technical Methodology

To assess the criteria presented above, we used a range of technical methods: decrypting and eavesdropping on network traffic, static analysis of the companion apps, network traffic evaluation, and Wi-Fi sniffing. Figure 1 gives an overview of our setup.

*Wi-Fi Sniffing.* Some toys communicate with their companion app or a local web server instead of (or in addition to) accessing the internet directly. To analyze this network traffic, we use a Raspberry Pi with `aircrack-ng` [37] and `Wireshark`. To filter out unrelated traffic, we specify the channel and BSSID of the local Wi-Fi. When the local Wi-Fi network uses WPA2 encryption with a password, `Wireshark` can decrypt the traffic after entering the raw WPA PSK key, which can be generated on the `Wireshark` website [64].

*Decryption of Network Traffic.* To intercept and decrypt TLS traffic between companion apps and the internet, we perform a Man-in-the-Middle (MitM) attack using `mitmproxy` [9]. To achieve this, the app must (1) trust the certificate of the MitM attacker and (2) not use a hardcoded certificate for the remote server (SSL pinning).

To ensure that apps trust the mitmproxy certificate, we install the certificate on the phone and transfer it from the user-trusted to the system-trusted certificates folder. We use a rooted phone to enable global installation of the mitmproxy certificate.

To bypass SSL pinning, we use `Frida`, a tool for dynamic code instrumentation [42]. Instead of injecting the `Frida` server into companion apps via app patching, we opted to run the `Frida` server directly on the phone. This approach allows us to easily execute scripts that modify the applications, to bypass their SSL pinning code [44]. Once more, we use our rooted phone to gain the superuser access necessary for this.

In the `mitmweb` user interface of `mitmproxy`, we can then examine the decrypted data and save it in an `har` (HTTP Archive) file for further analysis.

*Static Analysis.* We used static analysis based on the Mobile Security Framework `MobSF` [1] to analyze the requested permissions and trackers embedded in the toys' companion apps. `MobSF` provides vulnerability assessment, evaluates trackers and permissions, and summarizes metadata.

In addition to the automated `MobSF` evaluation, we also conducted a manual assessment by inspecting the apps' `Java` code. To navigate the large code bases efficiently, we searched for strings that reveal information about encoding (*encode, decode, base64, utf-8, H264, H265, png, gif, jpeg*), encryption and hashing (*encrypt, decrypt, TLS, AES, hash, SHA, MD5*), authentication (*password, pwd*), and the Google Ad ID (*advertise, gaid, ad_id*).

*Network Traffic Analysis.* To capture the network traffic from companion apps and toys, we use `Wireshark` [63]. For each toy, we start `Wireshark` data capture, activate the device and/or launch the companion app, log in if required, start a game, and perform typical user interactions for 2–10 min. Finally, we shut down all processes and end the data capture. A `python` script is used to filter and summarize the captured traffic. First, we filter out any traffic that does not involve the phone or toy, based on their IP addresses. Then, we aggregate incoming and outgoing messages to specific servers. We resolve server IP addresses to server names, hosting providers and geolocations using the `geolocation-db` [56] API and manual lookups [25]. To assess encryption with TLS, we parse the `Client Hello` messages to extract available TLS cipher suites. We cross-reference these suites with cryptographic security evaluations provided by an Ciphersuite Info API [48] and evaluate the negotiated TLS cipher suite. On the server side, we list available cipher suites using `SSL Labs` [45].

## 4    Results

In this section we present our findings regarding the cybersecurity, privacy, transparency and compliance of the 12 selected toys. Table 2 summarizes the results.

### 4.1    Security

*Communication.* For connections outside the local network to the internet, most toys establish secure communication channels to servers via TLS and HTTPS using secure TLS cipher suites, even though most toys still use TLS 1.2 instead of TLS 1.3 (Fig. 2). Although the Tigerbox encrypts traffic for most servers, traffic to the firmware update server remains unencrypted. While the update

**Table 2.** Overall results for each criterion and each toy. Red triangles (▲) indicate deficiencies, vulnerabilities, or inadequacies; yellow squares (■) indicate partial fulfillment of the criterion (more detail in the text); blue circles (●) indicate that the toy fulfills the criterion well. A dash (-) indicates that the criterion is not applicable to the toy.

| | Edurino | Kidibuzz | Camera | Moorebot | Osmo | Pictionary | Tamagotchi | Tigerbox | Tiptoi | Toniebox | Twister | Winky |
|---|---|---|---|---|---|---|---|---|---|---|---|---|
| **Security** | | | | | | | | | | | | |
| Global network traffic encrypted | ● | ● | - | ■ | ● | ● | ● | ● | ▲ | ■ | ● | ● |
| Local network encryption | - | - | ▲ | ■ | - | - | ■ | - | ▲ | ▲ | - | - |
| Wi-Fi setup | - | ● | - | ■ | - | - | ■ | ● | ▲ | ▲ | - | - |
| SW & HW support periods | ▲ | ▲ | ▲ | ▲ | ▲ | ▲ | ▲ | ▲ | ▲ | ▲ | ▲ | ▲ |
| **Privacy** | | | | | | | | | | | | |
| Processing minimum necessary | ▲ | ■ | ● | ■ | ▲ | ■ | ■ | ● | ● | ■ | ■ | ■ |
| Number of trackers | 1 | - | 0 | 2 | 1 | 3 | - | 1 | - | 4 | 2 | 0 |
| Location of server | ■ | ■ | - | ▲ | ■ | ■ | ■ | ■ | ● | ■ | ▲ | ■ |
| Only necessary app permissions | ● | - | ▲ | ▲ | ▲ | ■ | - | ● | - | ▲ | ■ | ■ |
| **Transparency** | | | | | | | | | | | | |
| Subject data access requests | ▲ | ▲ | - | ▲ | ■ | - | - | ● | - | ● | - | ▲ |
| Privacy policy: Transparent info about data | ● | ■ | ▲ | ▲ | ■ | ■ | ■ | ■ | ■ | ● | ● | ■ |
| Privacy policy: Consent and withdrawal | ● | ● | ▲ | ▲ | ● | ■ | ● | ● | ● | ● | ■ | ● |
| UI: Updates available | ● | ■ | ■ | ● | ● | ● | ● | ● | ● | ● | ● | ● |
| UI: Info about risks/disruptions | ■ | ▲ | ■ | ■ | ■ | ■ | ▲ | ● | ■ | ■ | ■ | ■ |
| UI: Model and version designation | ▲ | ● | ▲ | ■ | ▲ | ● | ● | ● | ■ | ● | ● | ■ |

itself is encrypted, reducing the risk of firmware exposure or malicious modification, it may allow for offline attacks against firmware updates or the encryption mechanism. Tiptoi also sends data over HTTP without encryption. Although the transmitted data mainly involves audio file downloads and updates, which may not contain sensitive information, passive observers may be able to profile usage of the toy.

Communication from the Moorebot robot appears to be RTMP encoded rather than encrypted, though decoding attempts were unsuccessful. The Toniebox encrypts data to servers using TLS, but uses a weak TLS cipher suite (using SHA1 and the AES in the Cipher Block Chaining (CBC) mode, which are considered insecure in TLS 1.3). The Toniebox server only supports cipher suites which are classified as "weak", whereas the Toniebox device supports "weak" and even worse, "insecure" cipher suites.

All local network connections initiated by the toy devices themselves are unencrypted. The Children's camera transmits images only WPA2 encrypted with the default and unchangeable password "12345678", which offers only limited protection. During initialization, Toniebox and Tiptoi send the user's Wi-Fi

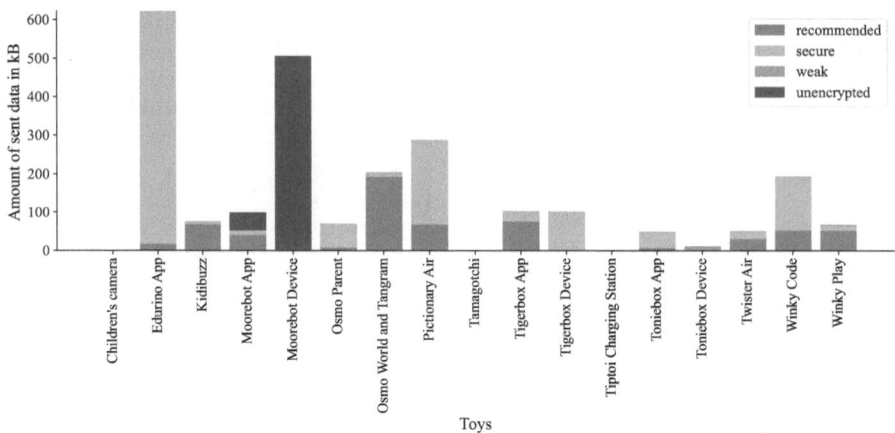

**Fig. 2.** Amount of data sent (in kB) for all toys, by security of the negotiated TLS cipher suite (weak/secure/recommended according to TLS 1.3 standard). Packets without payload (e.g. Acknowledgments) are excluded.

credentials over unencrypted HTTP, creating a security risk for disclosure as they are not even WPA2 encrypted.

*Wi-Fi Setup.* Some of the toys create their own Wi-Fi network for communication, while others ask for credentials to connect to an existing Wi-Fi network by the home's router. For IoT devices without a user interface, such as Moorebot, Tiptoi, Toniebox and the Children's camera, evaluation of the connection is more challenging. These toys set up their own local Wi-Fi network for sharing the Wi-Fi credentials to connect to the router.

Tiptoi and Toniebox have open Wi-Fi networks without password protection, allowing anyone to connect to the local network. Once connected, the user inputs their home Wi-Fi credentials on their phone or laptop, which are then transmitted to the device over HTTP without TLS encryption. With the Wi-Fi sniffing setup described earlier we are able to intercept the connection and can see the plaintext transmission of the given Wi-Fi credentials. This is concerning as it allows nearby attackers to retrieve the Wi-Fi credentials, especially since Tiptoi and Toniebox are popular toys in public libraries. The identified security vulnerabilities and proposed solutions were communicated to the respective vendors. Toniebox acknowledged the issue and stated that they are already working on a solution for future releases, whereas Tiptoi only indicated an intent to investigate the issue further.

The Moorebot robot provides its own WPA2 password-protected Wi-Fi network with the default password "r0123456". This password is the same for every robot. However, users can change the password when setting up the device. After connecting and entering the home Wi-Fi credentials, the credentials are transmitted to the device in encoded form. Then, users can choose to connect the robot to the companion app, via their home Wi-Fi or the device's Wi-Fi network.

The Tamagotchi toy offers two ways to connect to the user's home Wi-Fi. By entering the Wi-Fi credentials directly into the device using just three buttons, or by scanning a QR code. Through scanning the QR code, Tamagotchi opens a local network, to which the users connects to. The Wi-Fi password is "596deccf4e1b1c979ae2af8e", and unique for each device, making it harder for an attacker to crack the WPA2 encryption. However, the traffic itself is not encrypted and if the password is known, it is also vulnerable to malicious interception and disclosure of the Wi-Fi credentials.

The Winky robot connects to the phone and companion app with Bluetooth 5 Low Energy. It uses Secure Connections for pairing which provides protection against passive eavesdropping, however, its use of the Just Works association method leaves the connection open to Man-in-the-Middle attacks.

*HW/SW Support Periods.* No toy manufacturer declares hardware or software support periods for their products. The terms of service declarations primarily outline standard return rights and warranty claims, without explicit mention of ongoing hardware or software support. The vendors of three toys – Tamagotchi, Twister Air and Winky – even declare that they may discontinue services and support without notice. The Kidibuzz phone comes with Android 10 (released in 2019), which Google stopped supporting in March 2023. Concerningly, we bought the phone in October 2023 and it is still available as of April 2024.

### 4.2  Privacy

*Personal Data.* To evaluate the users' privacy, we analyze the decrypted traffic to see which kind of personal data is transmitted to which service. We also examine whether the privacy policy explains this data collection and processing.

For many toys, we find that the privacy policies do not state clearly which specific data is collected and for what purpose the vendors are processing it.

By examining the transmitted data and evaluating their necessity (see Table 3), we find that especially Edurino and Osmo, the learning apps, perform poorly. Edurino sends data about the child's sex, the created avatar, and detailed game analytics. Osmo asks for the full name and the child's exact date of birth. Another case of unnecessary data collection is done by Toniebox as they describe in their privacy policy that they send the list of all SSIDs in the area to their server. This is not needed for the functionality and may allow the vendor to geolocate the Toniebox and identify which Toniebox owners live nearby.

Almost every toy sends device and unique identifiers together with analytics data such as detailed interaction logs, session lengths and device information such as model, OS version and CPU details. While much of this data is justified by the vendors as necessary for improving the app and user experience, it may not be essential for the app's functionality. Moreover, collecting such data could result in the creation of user profiles and identifiable fingerprints.

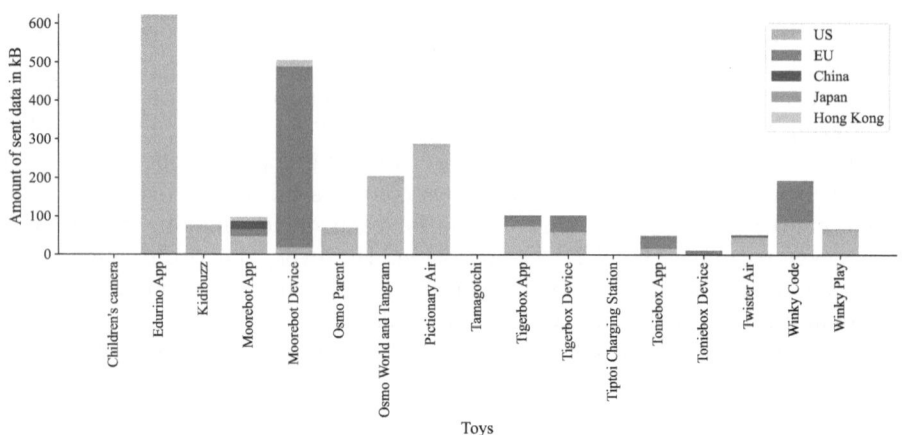

**Fig. 3.** Amount of sent data (in kB) for all toys, by location of the data destination. Packets without payload (e.g. Acknowledgments) are excluded.

Interestingly, we find code in most of the apps to retrieve the Google Advertisement ID. As we do not see the Ad ID directly in the traffic, it is uncertain whether the identifier is actually being transmitted to vendors or third-parties. However, given the functionality exists, it is possible that the Ad ID is scrambled into the User ID, as described in the Twister Air privacy policy. Alternatively, the functionality may be inactive, but could be turned on in a future version.

We also find various services which are identified as trackers in the code and in the captured traffic. Seven out of nine companion apps use trackers, Toniebox being the leader by employing four tracking services. Moreover, users should be aware that simply using these applications will result in data being sent to Google Libraries which we can see in the captured traffic. In Fig. 4, we provide a list of all contacted services, summing up all transmitted data by all toys.

An analysis of server locations shows that major hosting providers, such as Google Cloud and Amazon AWS, are mainly used, with servers primarily located in the United States. Two toys, the Moorebot and Twister Air, transmit data to servers located in China, specifically to the Alibaba Cloud. Only one toy, the Tiptoi pen, communicates entirely with servers in Europe. An overview of the server locations for each toy can be seen in Fig. 3.

In addition, on local networks the Moorebot has an open Robot Operating System (ROS) version 1.4 interface. ROS employs a publisher and subscriber architecture, allowing anyone on the same network as the robot to subscribe to topics. As a result, we were able to access the live video stream, audio stream and other sensor data without any security layer such as authentication or encryption. Furthermore, by publishing topics, we could control the robot's movements to capture additional video data from the environment or improve audio quality of the audio stream.

**Table 3.** Data types transmitted by each toy and assessment whether each data type is required to provide the toy's functionality. Red triangles (▲) indicate deficiencies, vulnerabilities, or inadequacies; yellow squares (■) indicate partial fulfillment of the criterion (more detail in the text); blue circles (●) indicate that the toy fulfills the criterion well. Cases where the evaluation is based on public data instead our traffic analysis are marked with an asterisk (*).

| | Edurino | Kidibuzz | Camera | Moorebot | Osmo | Pictionary | Tamagotchi | Tigerbox | Tiptoi | Toniebox | Twister | Winky |
|---|---|---|---|---|---|---|---|---|---|---|---|---|
| Processing minimum necessary | ▲ | ■ | ● | ■ | ▲ | ■ | ■ | ● | ● | ■ | ■ | ■ |
| **Personal Information** | | | | | | | | | | | | |
| Email | ● | ● | - | ● | ● | - | - | ● | - | ● | - | ● |
| Name/pseudonym | ■ | ▲ | - | - | ▲ | - | ● | ● | - | ● | ● | ▲ |
| Birthday/age | ■ | - | - | - | ▲ | - | ■ | ● | - | ● | - | - |
| Avatar | ▲ | - | - | - | ● | - | ▲ | ■ | - | ● | - | - |
| Sex | ▲ | - | - | - | - | - | - | ● | - | - | - | - |
| Region/location | ● | ● | ● | - | - | ● | ● | ● | - | ■ | ■ | ● |
| Wi-Fi credentials | - | ●* | - | ●* | - | - | ●* | - | ● | ▲ | - | - |
| **Identifiers** | | | | | | | | | | | | |
| User ID | ● | - | - | ● | - | ● | - | - | - | ● | ● | ● |
| Ad ID | ■* | - | - | ■* | ■* | ■* | - | - | - | ■* | ■* | - |
| Device ID | ■ | ■ | - | ■ | ■ | ■ | ● | - | - | ■ | ■ | ■ |
| **Game Analytics** | | | | | | | | | | | | |
| Device Data | ■ | ● | - | ● | ● | ■ | ● | - | - | ● | ● | ■ |
| Game Data | ■ | ●* | - | ●* | ■ | ■ | ■ | ■ | - | ■ | ■ | ■ |
| Interactions | ■ | ●* | - | ●* | ■ | ■ | - | - | - | ■ | ■ | ■ |
| Image data | - | ●* | - | ■* | - | ● | - | - | - | - | ● | ■ |
| Audio data | - | ●* | - | ■* | - | ● | - | ● | - | ● | ● | ■ |
| **Product Analytics** | | | | | | | | | | | | |
| Purchase information | ▲ | - | - | - | ▲ | - | - | - | - | - | - | - |

*App Permissions.* We extracted app permissions from the `AndroidManifest` file and evaluated them regarding necessity for the intended functionality (see Table 4). Positive examples are Edurino and Tigertones, the companion app for Tigerbox, that either require only minimal permissions overall or not very intrusive ones. Pictionary Air, Twister Air and both Winky companion apps perform overall well. However, permissions like the exact location, access to the phone status or the possibility to run at startup are seen as unnecessary.

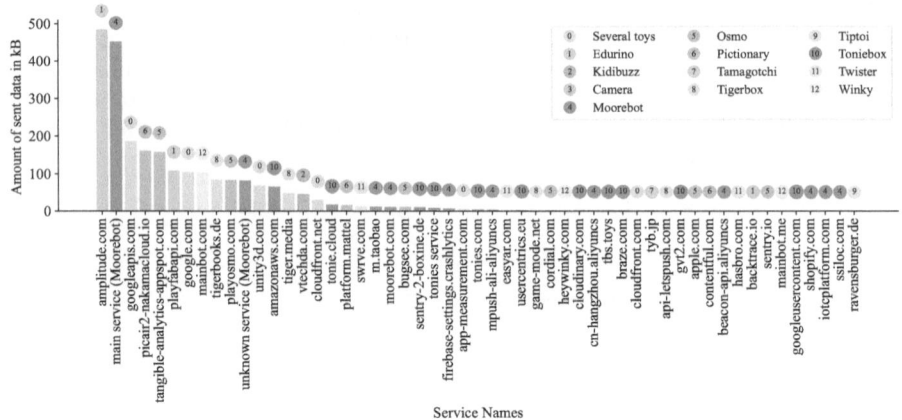

**Fig. 4.** Amount of data sent (in kB) across all services, by toy. Packets without payload (e.g. Acknowledgments) are excluded.

In contrast, the Children's camera, Moorebot, Osmo Parent and the Toniebox app request extensive permissions beyond what is necessary for their operation. This includes the phone status to see the device ID and call information, location status, reading external storage, accessing contact information, and the ability to run at startup.

## 4.3 Transparency

*Subject Data Access Requests.* We sent subject data access and deletion requests to all toy vendors for which we have a user account. Only three of seven vendors replied within the one-month period mandated by the GDPR.

Osmo answered quickly and confirmed deletion of the data. However, they indicated that they had only stored the e-mail address, even though we observed transmissions of birth dates and game data to their servers and synchronization with the Osmo parent app, indicating that more data is stored on their side. The data access process for the Toniebox is unnecessarily complicated: users have to provide either PGP keys for the e-mail address associated with their account, or their real-world home address. We requested the digital form and received an encrypted e-mail with a PDF containing the list of data they claim to have stored. However, the list does not contain scanned SSIDs or behavioral analytics, which are processed according to their privacy policy and confirmed to be linkable to user accounts [53]. Tigerbox responded with a list of all personal data, a history of stories and songs we listened to, as well as a breakdown of sent e-mails and whether they were opened or not.

*Privacy Policies.* In line with prior work on privacy policies [60], the privacy policies for our toys are on average too vague and unspecific, especially regarding transparent information about detail of user data processing, storage and

**Table 4.** App permissions requested by companion apps and evaluation whether the permissions are necessary for the intended functionality. Permissions with a △ sign are "dangerous" permissions according to Google [21]. Permission marked with an asterisk (*) are required for some features but not the main functionality of the app.

| | Edurino | Kidibuzz | Camera | Moorebot | Osmo | Pictionary | Tamagotchi | Tigerbox | Tiptoi | Toniebox | Twister | Winky |
|---|---|---|---|---|---|---|---|---|---|---|---|---|
| Only necessary app permissions | ● | - | ▲ | ▲ | ▲ | ▨ | - | ● | - | ▲ | ▨ | ▨ |
| Number of permissions | 3 | - | 10 | 19 | 15 | 18 | - | 21 | - | 32 | 14 | 8 |
| Location △ | - | - | ▲ | - | - | ▲ | - | - | - | ▲ | - | ▲ |
| Phone status △ | - | - | ▲ | ▲ | - | - | - | - | - | ▲ | ▲ | - |
| Contacts △ | - | - | - | ▲ | - | - | - | - | - | - | - | - |
| Run at startup | - | - | - | ▲ | - | - | - | - | - | - | ▲ | - |
| Camera △ | - | - | ● | - | ▲ | ● | - | - | - | ▲ | ● | - |
| Microphone △ | - | - | ▲ | ▲ | - | ●* | - | ●* | - | ●* | - | - |
| Read storage △ | - | - | ●* | ▲ | ▲ | ▲ | - | ●* | - | ●* | ▲ | - |
| Write storage △ | - | - | ●* | ●* | ▲ | ●* | - | ●* | - | ●* | ●* | - |
| Post notifications △ | - | - | - | ▲ | ▲ | ▲ | - | ▲ | - | ▲ | ▲ | - |
| Network connections | ● | - | ▨ | ● | - | - | - | - | - | ● | ● | ● |
| Wi-Fi | - | - | ● | ● | - | - | - | - | - | ● | - | - |
| Bluetooth △ | - | - | - | - | - | ●* | - | ●* | - | ●* | - | ● |
| NFC | - | - | - | - | - | - | - | ●* | - | - | - | - |

transmission (see Table 5). Most toys have easily accessible privacy policies. The Twister Air app stands out for its detailed list of data and purposes, while others, such as Pictionary Air, Children's camera, or Tigerbox, have generic privacy policies that mention the processing of personal data without specifying purposes and recipients.

The Moorebot robot lacks a privacy policy for its device entirely. Its user manual mentions data being sent to AWS servers in the US without specifying the data or its use. Finding a privacy policy for the Children's camera is challenging for users, as no mention is provided in the app or user manual. Only the link on the Apple App Store (but not the Google Play Store) leads to the privacy policy, which is very generic. While a privacy policy is missing from both the Winky Play app and the website, the Winky Code app has a good privacy policy that users have to consent to. Tiptoi provides a privacy policy during the setup of Wi-Fi credentials. Due to their lack of data collection, the policy is brief.

Obtaining consent for most apps and toys involves displaying the privacy policy and sometimes the terms of service during the initial setup process. In some cases, users may need to actively check a consent box. Tigerbox and Toniebox offer options for users to opt-in to receive advertisements via email, while Edurino allows users to decide whether their data should be stored in the Edurino cloud. However, Edurino does not provide any information about the location or hosting service of the stored data.

**Table 5.** Evaluation of the availability of privacy policies and their quality regarding specific information on what data is collected/shared, and for which purpose.

| | Edurino | Kidibuzz | Camera | Moorebot | Osmo | Pictionary | Tamagotchi | Tigerbox | Tiptoi | Toniebox | Twister | Winky |
|---|---|---|---|---|---|---|---|---|---|---|---|---|
| Transparent info about data protection | ● | ■ | ▲ | ▲ | ■ | ■ | ■ | ■ | ■ | ● | ● | ■ |
| Availability of privacy policy | ● | ● | ■ | ▲ | ● | ● | ● | ● | ■ | ● | ● | ■ |
| Which data | ● | ● | ■ | ▲ | ■ | ▲ | ● | ■ | ● | ● | ● | ● |
| How used | ■ | ■ | ▲ | ▲ | ■ | ▲ | ■ | ▲ | - | ■ | ● | ● |
| By whom | ● | ▲ | ▲ | ▲ | ■ | ▲ | ■ | ▲ | - | ● | ● | ● |

*User Interface Information.* The presentation of privacy- and security-relevant information to users could be improved. For example, none of the toys offer a password strength indicator that would help users create stronger passwords. In addition, the Kidibuzz fails to explain all password requirements when rejecting passwords, such as the prohibition of using names or dictionary words.

The provision of information about software updates is also inadequate. Most IoT devices either lack descriptions of updates or provide vague details, often omitting descriptions of security fixes. One exception is the Tigerbox, which provides a changelog for updates to both device and its companion app.

Furthermore, not all version and model numbers are available to users. While app versions can be viewed in the settings of an Android phone, firmware version information is provided by the toy manufacturer. Unfortunately, this information is not available for the Toniebox, and model numbers are missing for the Tiptoi pen and the Moorebot.

### 4.4   Compliance

*General Data Protection Regulation (GDPR).* The GDPR mandates vendors to ensure a basic level of protection mechanism for user's data and control mechanisms such as the availability of privacy policies, data minimization, provision of secure storage, and data retention rules. With a lens of technical, but not legal, expertise, we estimate that not all toys are compliant with the GDPR. In particular, we find shortcomings in the availability of the privacy policy (three vendors) and in the detail provided in these policies about specific data processing practices. Many vendors do not list the data items they are processing, or omit information about which parties the data is shared with.

Consumers also have the right to access their data, ask for correction and for deletion. Most vendors mention this right in their privacy policy, however, the response rate of 43% for our subject access requests is disappointing.

*Cyber Resilience Act (CRA).* The Cyber Resilience Act [18], approved by the European Parliament in March 2024, seeks to establish a standardized cybersecurity framework for products and software with digital components. The CRA, in the current version, mandates for software updates and vulnerability disclosure rules, requiring regular updates with transparent documentation of mitigated vulnerabilities and detailed software bills, listing the history of all detected and patched vulnerabilities. In addition, the CRA asks for basic cybersecurity requirements such as secure authentication or to minimize the attack surface. It will also ask vendors to publish vulnerability disclosure policies, offering security researchers or attentive users a guideline how to disclose found vulnerabilities.

In our examination, we see that only three of 12 vendors provide this information: Tonies (Toniebox), Hasbro (Pictionary Air) and VTech (Kidibuzz). Furthermore, regarding information on software updates, no vendor with the exception of Tigerbox provides detailed information on updates or a history log. This means that most vendors will need to update their processes to comply with the CRA, once it comes into effect.

### 4.5   Limitations

Limitations of this work include the small sample size of only 12 toys selected from a much larger smart toy market, which may not be representative for the whole product range.

We did not conduct a longitudinal study, which prevents us from observations of any potential changes in data collection practices over time, and restricts us in assessing long-term effects.

Furthermore, we did not attempt to extract or reverse engineering device firmwares. Thus, plaintext traffic between toys and internet could not be analyzed as we could not install trusted certificates for Man-in-the-Middle attacks.

Finally, we focused our analysis of compliance on two recent and coming pieces of legislation (GDPR and CRA), but did not analyze other relevant legislation such as the ePrivacy directive.

## 5   Discussion and Recommendations

In this section we discuss the limitations of our study, put our findings regarding cybersecurity, privacy, transparency and compliance to EU legislation in a broader context, and give specific recommendations for improving smart toys.

## 5.1   Security

In our examination of current communication practices, we see the widespread use of TLS 1.2 encryption for connections to external servers. While some cipher suites within TLS 1.2 offer forward secrecy, their use in our examination, as well as not using the newer TLS 1.3, is not best practice. Most toys as well as the corresponding servers support the stronger, forward secrecy enabling algorithms, such that an upgrade is theoretically possible. Only Toniebox relies on a weak cipher suite for communication, which offers an attack surface to decrypt the traffic.

We also identified some instances of unencrypted connections, which compromises data confidentiality and undermines the integrity and authenticity of transmitted information. We see plaintext traffic mostly in local networks, for example in the Wi-Fi setup procedure, where Wi-Fi credentials can be easily obtained by eavesdropping on the connection. Although the probability of an external attack on the local network is low, these devices are popular in public libraries, which creates a scenario where attackers could exploit the setup process by following users home to capture their Wi-Fi credentials. As this issue is present for all IoT devices without a user interface, the new Wi-Fi Device Provisioning Protocol (DPP, or "Easy Connect") offers a secure solution for connecting IoT devices to Wi-Fi networks [62]. In that solution, devices can be onboarded to a network through QR codes, NFC tags or user-chosen configurators, with strong encryption. None of our toys has deployed this protocol.

Additionally, vendors fail in disclosing information on their device's support periods. This is concerning because users are not informed when security updates are discontinued, and devices may cease functioning entirely.

## 5.2   Privacy

Personal data has become a prime target for both commercial interests and malicious actors. Children represent an even more vulnerable group in the digital landscape, particularly due to their limited understanding of the implications of sharing personal information online. Moreover, unauthorized access to children's personal data creates a number of potential risks, including exposure to inappropriate content, manipulation through targeted advertising and creating digital footprints, potentially influencing the child's future.

We observe extensive collection of game analytics and behavioral data, including time spent on games, hints/errors, and toy interactions. This data has the potential to generate user-specific profiles, reflecting children's intelligence, problem-solving abilities, and even daily behavioral patterns like playtime and bedtime routines.

Another concerning observation is the widespread use of data tracking mechanisms. These trackers collect user data to create profiles for targeted advertising or for sale to third-party data brokers [27].

Furthermore, we observe network traffic directed to servers in the US, Europe, Japan and China. Although the European Union has established privacy treaties with the US and Japan [14] to protect the data of European citizens [15], data protection concerns remain about the adequacy of privacy measures outside of the EU and to the US [2,40]. In particular, China is not regarded as having an adequate level of data protection [15], which means that children's interactions with their toys may be disclosed to Chinese authorities and other third parties.

Not all apps comply with the best-practice to only request minimum necessary app permissions. Modern Android apps often ask users to grant permissions during runtime, including for access to location, Bluetooth, camera, and microphone. However, not all users – which may well be children, in the case of toy companion apps – may understand which permissions are really needed, and therefore may not be able to make informed security and privacy choices [30].

### 5.3 Transparency

The process and responses of the accessing the own personal data stored by the vendor is insufficiently implemented, which is in line with prior work on subject access requests [57]. Only 43% of vendors responded within the one-month period, and their responses did not include all transmitted data that we could see in our data collection. Furthermore, the process of requesting data is too cumbersome: users need to write emails, send PGP keys, or provide their home address. In the latter case, many users may not have PGP keys (or may be unable to use PGP, which is a well-known issue [61]), and would be left with a paper-based process that requires them to reveal additional data (their address) to the vendor. This is clearly not ideal. A better solution would be to allow users to download their data directly in the app. This would not only be more user friendly, but the transmission could also be encrypted with TLS. However, this approach is not implemented by any of the vendors in our study.

Our analysis finds that many toy companies fail to provide clear and transparent privacy policies. The inconsistency in the availability and detail of privacy policies is evident, with some toys providing detailed lists of data and purposes, while others offer only generic statements. This leaves users unaware regarding the use of their personal information, which is particularly concerning given that sensitive data of children may be processed.

The process for providing and withdrawing consent is insufficient overall. While some toys offer check boxes for proactive consent in the set-up process, others may not present the privacy policy at all, or only refer to it in the settings menu. In addition, the ability to revoke consent is limited. Providers typically do not offer granular choices for data processing, resulting in a binary "all or nothing" scenario. This lack of flexibility may force users to consent to data processing practices that they would otherwise prefer to avoid [28].

### 5.4 Compliance

Our results show that some vendors fall short of full GDPR compliance. In particular, we find shortcomings in the availability and comprehensiveness of privacy policies. Even when provided, these policies often lack the detailed information necessary to adequately inform users about their data processing practices. Furthermore, we see that not all vendors apply data minimization practices as we observe large amounts of game analytics and device information data. Moreover, the right to access user data and the right to data deletion is ignored by 57% of vendors, resulting in non-compliance with the GDPR.

The CRA aims to improve cyber resilience for products with digital components. Thus, a focus lies on other cybersecurity metrics, primarily regarding patching of vulnerabilities. The act will enforce vulnerability disclosure policies and better information on software updates, such as providing a software bill with the update history. Currently, only one vendor provides a changelog of software updates, and three vendors provide vulnerability disclosure policies, indicating that most vendors need to catch up on the requirements of the CRA. This means that in terms of transparency on patched security vulnerabilities users can hope for better information after the adoption of the CRA. However, the success of this legislation also relies on the enforcement practices, as the descriptions and requirements are held very general.

## 5.5  Recommendations

Upon the discussion of our findings we want to provide some recommendations regarding the cybersecurity, privacy and transparency of children's smart toys.

*Prioritizing Privacy and Security.* Toy vendors have a particular responsibility, due to their young target audience, to implement security-by-design and privacy-by-design best practices throughout their development lifecycle. This includes strong encryption, proper authentication mechanisms, data minimization, and the provision of transparent information and choices to users.

*Use of a label system.* The creation of a unified label inspired by the nutrition label model [12] could inform users on first glance about basic security, privacy and transparency features of the toy. This label should be placed on the packaging and in online stores for easy accessibility. Possibly, this label could motivate vendors to invest and develop more secure and privacy-preserving smart toys.

*Enhance User Control.* User interfaces should provide more easily accessible information about the data collected by the vendor. This should be combined with options to withdraw consent and delete specific data as desired.

*Easier Subject Data Access Requests.* Users should be able to submit subject data access requests by means of a simple button press within the app. Automated processing of these requests on the vendor-side, i.e., giving users a direct download, would significantly improve the user experience.

*Opt-in Consent.* Toys should rely more on opt-in rather than opt-out approaches to gain consent from users. This ensures that users need to actively choose to share data and avoids excessive data collection from users who did not change the default settings.

*Enable Fine-Grained Consent.* Users should be able to selectively disagree with specific aspects of a privacy policy, giving users more granular control. For example, this should include options for users to refuse collection of behavioral and game analytics data.

## 6    Conclusion and Future Work

We examined 12 children's smart toys with respect to cybersecurity, privacy, transparency and compliance to EU legislation and formulated recommendations aimed at improving security and privacy for smart toys.

Our investigation of security issues uncovered that some connections especially in local networks are inadequately encrypted, leaving vulnerabilities to data disclosures. Furthermore, while the data transmitted is not excessively intrusive, the substantial volume of data, particularly toy analytics data combined with unique identifiers, raises privacy concerns, because it allows pervasive behavioral profiling and fingerprinting of children.

In addition, the toys do not provide sufficient transparency regarding the data being transmitted and its recipients because the privacy policies lack the desired detail. The process for accessing information through subject data access requests requires more user effort than necessary, and the low response rate indicates that not all toys are compliant with the GDPR.

Looking ahead, the CRA may address some of our security concerns, particularly in areas such as vulnerability patching. However, the effectiveness of this legislation will depend on its enforcement and the cooperation of toy vendors.

*Future Work.* In the future, we plan to deepen our study of smart toys, including data from the IoT devices themselves rather than just the companion apps. This includes a comprehensive examination of the types of data collected and transmitted, with a particular focus on understanding the implications of AI-powered and medical smart toys which have recently become available. These efforts may not only inform future regulatory frameworks, but may also be helpful for parents who decide whether or not to purchase a smart toy.

**Disclosure of Interests.** The authors have no competing interests to declare that are relevant to the content of this article.

## References

1. Abraham, A., Magofei, Dobrushin, M., Nadal, V.: Mobile Security Framework - Documentation. https://mobsf.github.io/docs/. Accessed Dec 2023
2. Batlle, S., van Waeyenberge, A.: EU-US data privacy framework: a first legal assessment. Europ. J. Risk Regul. **15**(1), 191–200 (2024)
3. Bundesnetzagentur: Bundesnetzagentur zieht Kinderpuppe Cayla aus dem Verkehr. https://www.bundesnetzagentur.de/SharedDocs/Pressemitteilungen/DE/2017/14012017_cayla.html (Feb 2017). Accessed Oct 2023

4. Buttarelli, G.: Privacy matters: updating human rights for the digital society. Heal. Technol. **7**(4), 325–328 (2017). https://doi.org/10.1007/s12553-017-0198-y

5. de Carvalho, L.G., Eler, M.M.: Security tests for smart toys. In: ICEIS (2), pp. 111–120 (2018)

6. Castelluccia, C., Cunche, M., Le Métayer, D., Morel, V.: Enhancing transparency and consent in the IoT. In: 2018 IEEE European Symposium on Security and Privacy Workshops (EuroS&PW), pp. 116–119. IEEE (2018)

7. Chowdhury, W.: Toys that talk to strangers: a look at the privacy policies of connected toys. In: Arai, K., Bhatia, R., Kapoor, S. (eds.) Proceedings of the Future Technologies Conference (FTC) 2018: Volume 1, pp. 152–158. Springer International Publishing, Cham (2019). https://doi.org/10.1007/978-3-030-02686-8_12

8. Chu, G., Apthorpe, N., Feamster, N.: Security and privacy analyses of internet of things children's toys. IEEE Internet Things J. **6**(1), 978–985 (2018)

9. cortesi, maximilianhils, raumfresser: mitmproxy - an interactive HTTPS proxy. https://mitmproxy.org/. Accessed Oct 2023

10. Curio Interactive Inc: Homepage of the curio AI toy. https://heycurio.com (2024). Accessed Mar 2024

11. Emami-Naeini, P., Agarwal, Y., Cranor, L.F., Hibshi, H.: Ask the experts: What should be on an IoT privacy and security label? In: 2020 IEEE Symposium on Security and Privacy (SP), pp. 447–464. IEEE (2020)

12. Emami-Naeini, P., Dheenadhayalan, J., Agarwal, Y., Cranor, L.F.: An informative security and privacy "Nutrition" label for internet of things devices. IEEE Security Priv. **20**(2), 31–39 (2022). https://doi.org/10.1109/MSEC.2021.3132398

13. Escher, S., Weller, B., Köpsell, S., Strufe, T.: Towards transparency in the Internet of Things. In: Privacy Technologies and Policy: 8th Annual Privacy Forum, APF 2020, Lisbon, Portugal, October 22–23, 2020, Proceedings 8, pp. 186–200. Springer (2020). https://doi.org/10.1007/978-3-030-55196-4_11

14. European Commission: EU and Japan conclude landmark deal on cross-border data flows at High-Level Economic Dialogue. https://ec.europa.eu/commission/presscorner/detail/en/ip_23_5378 (2023). Accessed Apr 2024

15. European Commission: Adequacy decisions - How the EU determines if a non-EU country has an adequate level of data protection. https://commission.europa.eu/law/law-topic/data-protection/international-dimension-data-protection/adequacy-decisions_en (2024). Accessed Apr 2024

16. European Telecommunications Standards Institute: ETSI EN 303 645 V2.1: Cyber Security for Consumer Internet of Things: Baseline Requirements. https://www.etsi.org/deliver/etsi_en/303600_303699/303645/02.01.01_60/en_303645v020101p.pdf (2020). Accessed Oct 2023

17. European Union: General Data Protection Regulation. https://eur-lex.europa.eu/legal-content/EN/TXT/PDF/?uri=CELEX:32016R0679 (2016). Accessed Oct 2023

18. European Union: Cyber Resilience Act, preliminary version. https://eur-lex.europa.eu/legal-content/EN/TXT/PDF/?uri=CELEX:52022PC0454 (2022). Accessed Mar 2024

19. Fox, G., Lynn, T., Rosati, P.: Enhancing consumer perceptions of privacy and trust: a GDPR label perspective. Inform. Technol. People **35**(8), 181–204 (2022). https://doi.org/10.1108/ITP-09-2021-0706

20. Frolov, N.: An educational robot security research. https://securelist.com/smart-robot-security-research/111938/ (Feb 2024). Accessed Mar 2024

21. Google for Developers: Manifest.permission. https://developer.android.com/reference/android/Manifest.permission. Accessed Apr 2024
22. Hessel, S., Rebmann, A.: Regulation of Internet-of-Things cybersecurity in Europe and Germany as exemplified by devices for children. Int. Cybersec. Law Rev. **1**, 27–37 (2020). https://doi.org/10.1365/s43439-020-00006-3
23. Holland, M.: ChatGPT & Co.: In fünf Jahren personalisierte Gutenachtgeschichten vom KI-Teddy? https://www.heise.de/news/ChatGPT-Co-Tedybaeren-koennten-personalisierte-Gutenachtgeschichten-erzaehlen-9191143.html (2023). Accessed Mar 2024
24. Hung, P., Fantinato, M., Rafferty, L.: A study of privacy requirements for smart toys. In: PACIS 2016 Proceedings. Chiayi, Taiwan (Jun 2016). https://aisel.aisnet.org/pacis2016/71
25. IPSHU: IP Address Lookup Tools. https://en.ipshu.com/. Accessed Feb 2024
26. Johansen, J., et al.: A multidisciplinary definition of privacy labels. Inform. Comput. Secur. **30**(3), 452–469 (2022). https://doi.org/10.1108/ICS-06-2021-0080
27. Kollnig, K., et al.: Before and after GDPR: tracking in mobile apps. arXiv preprint arXiv:2112.11117 (2021)
28. Kollnig, K., et al.: A fait accompli? an empirical study into the absence of consent to Third-Party Tracking in Android Apps. In: Seventeenth Symposium on Usable Privacy and Security (SOUPS 2021), pp. 181–196 (2021)
29. Kotrba, K.: Alternative zu Tonies and Co: Lösen Startups mit KI-generierten Geschichten jetzt die Platzhirsche ab? https://www.businessinsider.de/gruenderszene/technologie/alternative-zu-tonies-co-loesen-startups-mit-ki-generierten-geschichten-jetzt-die-platzhirsche-ab/ (2024). Accessed Mar 2024
30. Kreuter, F., Haas, G.C., Keusch, F., Bähr, S., Trappmann, M.: Collecting survey and smartphone sensor data with an app: opportunities and challenges around privacy and informed consent. Soc. Sci. Comput. Rev. **38**(5), 533–549 (2020)
31. Macenaite, M.: From universal towards child-specific protection of the right to privacy online: Dilemmas in the EU general data protection regulation. New Media Society **19**(5), 765–779 (2017)
32. Manta, I.D., Olson, D.S.: Hello Barbie: First they will monitor you, then they will discriminate against you - perfectly. Alabama Law Rev. **67**(1), 135–188 (2015/2016). https://heinonline.org/HOL/P?h=hein.journals/bamalr67&i=145
33. Mattel: Shoppingpage Pictionary vs KI. https://shopping.mattel.com/de-de/products/pictionary-vs-ki-hyh74-de-de (2024). Accessed Mar 2024
34. McReynolds, E., Hubbard, S., Lau, T., Saraf, A., Cakmak, M., Roesner, F.: Toys that listen: a study of parents, children, and internet-connected toys. In: Proceedings of the 2017 CHI Conference on Human Factors in Computing Systems, pp. 5197–5207 (2017)
35. Mosenia, A., Jha, N.K.: A comprehensive study of security of internet-of-things. IEEE Trans. Emerg. Top. Comput. **5**(4), 586–602 (2016)
36. Nan, Y., et al.: Are you spying on me? Large-Scale Analysis on {IoT} data exposure through companion apps. In: 32nd USENIX Security Symposium (USENIX Security 23), pp. 6665–6682 (2023)
37. aircrack ng: Aircrack-ng. https://github.com/aircrack-ng/aircrack-ng. Accessed Feb 2024
38. NIST: Submission Requirements and Evaluation Criteria for the Lightweight Cryptography Standardization Process (2018). https://csrc.nist.gov/projects/lightweight-cryptography
39. Norval, C., Singh, J.: A room with an overview: towards meaningful transparency for the consumer internet of things. IEEE Internet of Things Journal (2023)

40. NOYB – European Center for Digital Rights: New Trans-Atlantic Data Privacy Framework largely a copy of "Privacy Shield". https://noyb.eu/en/european-commission-gives-eu-us-data-transfers-third-round-cjeu (2023). Accessed Apr 2024

41. OConnor, T., Jessee, D., Campos, D.: Through the spyglass: towards iot companion app man-in-the-middle attacks. In: Proceedings of the 14th Cyber Security Experimentation and Test Workshop, pp. 58–62 (2021)

42. oleavr: frida, Dynamic instrumentation toolkit for developers, reverse-engineers, and security researchers. https://frida.re/docs/android/. Accessed Oct 2023

43. Paracha, M.T., Dubois, D.J., Vallina-Rodriguez, N., Choffnes, D.: IoTLS: understanding TLS usage in consumer IoT devices. In: Proceedings of the 21st ACM Internet Measurement Conference, pp. 165–178 (2021)

44. pcipolloni: Frida Codeshare - Universal Android SSL Pinning Bypass with Frida. https://codeshare.frida.re/@pcipolloni/universal-android-ssl-pinning-bypass-with-frida/. Accessed Mar 2024

45. Qualys Inc: Qualys SSL Lab. https://www.ssllabs.com/ssltest/index.html. Accessed Feb 2024

46. Railean, A., Reinhardt, D.: OnLITE: on-line label for Iot transparency enhancement. In: Secure IT Systems: 25th Nordic Conference, NordSec 2020, Virtual Event, November 23–24, 2020, Proceedings 25. pp. 229–245. Springer (2021)

47. Ren, J., Dubois, D.J., Choffnes, D., Mandalari, A.M., Kolcun, R., Haddadi, H.: Information exposure from consumer Iot devices: a multidimensional, network-informed measurement approach. In: Proceedings of the Internet Measurement Conference, pp. 267–279 (2019)

48. Rudolph, H.C., Grundmann, N.: TLS Ciphersuite Search. https://ciphersuite.info/. Accessed Jan 2024

49. Sadeeq, M.A., Zeebaree, S.R., Qashi, R., Ahmed, S.H., Jacksi, K.: Internet of Things security: a survey. In: 2018 International Conference on Advanced Science and Engineering (ICOASE), pp. 162–166. IEEE (2018)

50. Schmidt, D., Tagliaro, C., Borgolte, K., Lindorfer, M.: IoTFlow: inferring Iot device behavior at scale through static mobile companion app analysis. In: Proceedings of the 2023 ACM SIGSAC Conference on Computer and Communications Security, pp. 681–695 (2023)

51. Shasha, S., Mahmoud, M., Mannan, M., Youssef, A.: Playing with danger: a taxonomy and evaluation of threats to smart toys. IEEE Internet Things J. **6**, 2986–3002 (2018)

52. Streiff, J., Noah, N., Das, S.: A call for a new privacy and security regime for Iot smart toys. In: 2022 IEEE Conference on Dependable and Secure Computing (DSC), pp. 1–8. IEEE (2022)

53. Team RevvoX: Toniebox Reverse Engineering. https://github.com/toniebox-reverse-engineering/talks/blob/master/2023-12-27%20-%2037C3%20-%20Toniebox%20Reverse%20Engineering.pdf and https://media.ccc.de/v/37c3-11993-toniebox_reverse_engineering#t=3609. Accessed Jan 2024

54. Trimananda, R., Le, H., Cui, H., Ho, J.T., Shuba, A., Markopoulou, A.: OVRseen: Auditing network traffic and privacy policies in oculus VR. In: 31st USENIX security symposium (USENIX security 22), pp. 3789–3806 (2022)

55. United Nations, G.: Convention on the Rights of the Child. https://www.ohchr.org/en/instruments-mechanisms/instruments/convention-rights-child (1989). Accessed Mar 2024

56. Unknown: Geolocation DB. https://geolocation-db.com/jsonp/. Accessed Feb 2024

57. Urban, T., Tatang, D., Degeling, M., Holz, T., Pohlmann, N.: A study on subject data access in online advertising after the GDPR. In: Pérez-Solà, C., Navarro-Arribas, G., Biryukov, A., Garcia-Alfaro, J. (eds.) Data Privacy Management, Cryptocurrencies and Blockchain Technology, pp. 61–79. Lecture Notes in Computer Science, Springer International Publishing, Cham (2019). https://doi.org/10.1007/978-3-030-31500-9_5

58. Wachter, S.: Privacy: primus inter pares - privacy as a precondition for self-development, personal fulfilment and the free enjoyment of fundamental human rights. SSRN Scholarly Paper ID 2903514, Social Science Research Network, Rochester, NY (Jan 2017)

59. Wachter, S.: The GDPR and the Internet of Things: a three-step transparency model. Law Innov. Technol. **10**(2), 266–294 (2018)

60. Wagner, I.: privacy policies across the ages: content of privacy policies 1996–2021. ACM Trans. Privacy Security **26**(3), 32:1–32:32 (2023). https://doi.org/10.1145/3590152

61. Whitten, A., Tygar, J.D.: Why Johnny can't encrypt: a usability evaluation of PGP 5.0. In: Usenix Security (1999)

62. Wi-Fi Alliance, Harkins, D.: Wi-Fi Easy Connect: simple and secure onboarding for IoT. https://www.wi-fi.org/beacon/dan-harkins/wi-fi-easy-connect-simple-and-secure-onboarding-for-iot (2023)

63. Wireshark Foundation: Wireshark - Network Protocol Analyzer. https://www.wireshark.org/. Accessed Dec 2023

64. Wireshark Foundation: WPA PSK (Raw Key) Generator. https://www.wireshark.org/tools/wpa-psk.html. Accessed Feb 2024

# Implementing ISO/IEC TS 27560:2023 Consent Records and Receipts for GDPR and DGA

Harshvardhan J. Pandit[1,2]([envelope]) [ORCID], Jan Lindquist[3], and Georg P. Krog[4]

[1] ADAPT Centre, Dublin City University, Dublin, Ireland
me@harshp.com
[2] Cybersecurity and Data Protection Group, National Standards Institute,
Dublin, Ireland
[3] Privacy and Security Group, Institute for Standards, Stockholm, Sweden
jan@linaltec.com
[4] Signatu AS, Oslo, Norway
georg@signatu.com

**Abstract.** The ISO/IEC TS 27560:2023 Privacy technologies—Consent record information structure provides guidance for the creation and maintenance of records regarding consent as machine-readable information. It also provides guidance on the use of this information to exchange such records between entities in the form of 'receipts'. In this article, we compare requirements regarding consent between ISO/IEC TS 27560:2023, ISO/IEC 29184:2020 Privacy Notices, and the EU's General Data Protection Regulation (GDPR) to show how these standards can be used to support GDPR compliance. We then use the Data Privacy Vocabulary (DPV) to implement ISO/IEC TS 27560:2023 and create interoperable consent records and receipts. We also discuss how this work benefits the implementation of EU Data Governance Act (DGA), specifically for machine-readable consent forms.

**Keywords:** consent · consent receipt · GDPR · DGA · ISO · Semantics

## 1 Introduction

*Informed Consent* is an important legal basis as it provides control and empowerment to data subjects or users based on the ability to choose and make decisions. Privacy and data protection laws such as EU's GDPR [16] regulate this process by defining conditions for when consent should be considered *Valid Consent*. The process of *Informed Consent* requires information be provided in the form of a *Consent Notice* to inform the data subject about the processing that will occur based on the consent and to enable them to make an *informed choice or decision*.

In order to assess whether an instance of given consent is valid thus requires keeping records of information regarding how the consent was obtained i.e. using

M. Jensen et al. (Eds.): APF 2024, LNCS 14831, pp. 228–251, 2024.
https://doi.org/10.1007/978-3-031-68024-3_12

the notice, and how the consent is being utilised i.e. the processing enabled through that consent. This same information is also required for the organisation to determine whether its processing activities should continue, e.g. depending on whether a particular user has given consent and whether it is still valid i.e. hasn't expired or wasn't withdrawn). Such information that is documented and maintained regarding consent is called a *Consent Record*.

ISO/IEC TS 27560:2023 Consent record information structure [6] is a Technical Specification that "specifies an interoperable, open and extensible information structure" for recording the data subject's consent to processing of their personal data i.e. as consent records, and to provide this information i.e. as consent receipts. The specification lists *information fields* that represent specific information associated with consent, and requirements over the form this information can take e.g. format, number of values, and whether it is mandatory or optional. It complements the earlier ISO/IEC 29184:2020 Online privacy notices and consent [5] which describes the information to be provided within privacy notices.

A ISO-27560 conformant implementation fulfils requirements by either storing information directly in the form prescribed by ISO-27560 or by storing information in a form that can be used to obtain this information. ISO-27560 allows flexibility in how the fields are represented to suit and match domain-specific labels or descriptions, or to introduce additional fields or information types that are needed. Such changes, expressed as *schemas* or *profiles*, are still required to be compatible with the requirements of ISO-27560 such as by requiring the same fields to be mandatory. In this manner, ISO-27560 provides a common, interoperable, and extensible structure for the exchange of information associated with consent.

In this article, we present an analysis of the requirements for recording consent within ISO-27560 and ISO-29184 and compare them with the requirements for valid consent under GDPR (Sect. 3). We then present our work in implementing ISO-27560 using the Data Privacy Vocabulary (DPV) [13,15] to create a machine-readable, interoperable, and extensible specification for consent records and receipts based on open standards (Sect. 4). Through this work we demonstrate the applicability and usefulness of ISO-27560 in assisting with the obligation for demonstrating consent under GDPR (Art.7-1), and explore how ISO-27560 and ISO-29184 can work within the legal framework of GDPR and DGA and the possibility for using this standard to inform the implementation of machine-readable common consent forms under the DGA (Sect. 5). We also discuss practicalities for implementations regarding trust and security (Sect. 6.1) and using records and receipts with eIDAS and EUDI wallets (Sect. 6.2).

## 2   Overview of ISO/IEC TS 27560:2023

**Goals and Scope.** ISO-27560 has two broad goals: to guide the recording of information about consent for processing of personal data in a form that is interoperable, open, and extendable, and to provide information to individuals. To implement this, it defines several requirements (as *controls* in ISO terminology) for ensuring the required information is maintained and is supported by appropriate processes within the organisation. ISO-27560 is stated as a supplement to the earlier ISO-29184, where ISO-29184 defines how information is provided via notices in order to request consent, and ISO-27560 defines how information is recorded for given consent and provided back to the individual (as receipts).

The objective of ISO-27560 is to define an information structure for consent record which contains: (1) Information about the processing of personal data; (2) Privacy notices where this information was provided; (3) How data was obtained; and (4) Events related to consent (giving, withdrawing, etc.). It also defines an information structure providing all or some of this information to the data subject in the form of a consent receipt. To support implementations, Annex A provides examples of consent records and receipts using DPV, and Annex B provides an overview of the different states or stages in 'consent lifecycle' - which is based on DPV's consent states [13, 15] and analysis of existing approaches [2, 8].

Specific guidance on implementation such as the choice of technologies is not in the scope of ISO-27560, though its Annexes provide informative guidance on related topics. Annex C describes performance and efficiency considerations, Annex D describes format and encoding structures, Annex E describes security of records and receipts, and Annex G describes application in Privacy Information Management Systems (PIMS). Further uses of consent records or receipts, such as how data subjects can obtain consent receipts or maintain their own consent records is not described in ISO-27560.

**Consent Records.** ISO-27560 defines *Consent Record* as the documentation of information about a data subject's consent for the processing of their personal data in terms of the details about the processing as well as the interactions related to consent (e.g. giving or withdrawing it). Consent Records are an essential part of keeping records regarding whether consent has been obtained and is valid for processing, and to keep this information for correctly conducting processing relying on it. ISO-27560 as well as regulatory requirements such as GDPR Article 7-1 require maintaining consent records where consent is used as the legal basis. While GDPR Article 7-1 only states that consent should be demonstrable, ISO-27560 provides an information structure for how this information should be maintained and what processes should exist within an organisation in for this.

It is important to distinguish between a *Consent Record* with several relevant but distinct concepts. A consent record only refers to the information recorded regarding consent, whereas a *Consent Notice* refers to the notice and information provided to the data subject in order to inform them about the processing - such as while requesting consent. While there is a significant overlap between a consent record and a consent notice, there are key differences such as notices orienting information for human consumption (e.g. layering of dialogues to provide summaries and detailed descriptions) and dictating the manner in which consent is expressed (e.g. checkboxes for options and confirmation by clicking a button). In contrast, a consent record is not required to accurately reflect the manner in which this information was presented to the user, but to only record it in a manner that enables assessing whether the consent is given and if so for which processing activities.

This distinction is evident in ISO-29184 being the standard for consent notices - which specifies what information should be present in a notice and the manner in which it is presented. In turn, ISO-27560 only refers to notices to limit its scope to representing information necessary within a consent record. Therefore a consent record, despite containing a link to the notice, is not by itself sufficient to determine the *validity* of consent, and instead acts as the primary record containing information or links to information for conducting such assessments. Its primary purpose is therefore limited to supporting claims for who is the subject, who is the controller, what is the consent about (e.g. which purpose, what recipients), what is the state of consent (e.g. request, given, terminated), and where/when/how it occurred (e.g. accepted on specific timestamp).

A ISO-27560 conformant consent record typically has the following sections representing relevant information:

1. metadata about the consent record such as its identifier
2. the individual associated with the consent i.e. data subject
3. the subject of consent i.e. specifics of the processing of personal data such as purposes, services, data categories, and storage conditions
4. entities involved e.g. data controller and third parties
5. relevant contextual information e.g. notice, rights, restrictions
6. provenance of events associated with the consent e.g. given, withdrawn

Under GDPR, the obligation to maintain records of consent is explicitly stated in Article 7-1 and Recital 42. This information includes, at a minimum, the identity of the Controller and the purposes of processing (Recital 42). Further, Articles 13 and 14 dictate the information that must be provided to the data subject which includes recipients, transfers to third countries, data storage periods and conditions, existence of rights (including consent withdrawal), and specific information regarding processing such as the use of automated decision making or profiling.

**Consent Receipts.** ISO-27560 defines a consent receipt as an authoritative document that is used to communicate the existence of a consent record or to provide information contained within it. It is effectively an 'authoritative copy' of a consent record provided by one entity to another, where it may contain all or only some information from the consent record regarding the consent and its relevance to processing activities. Such receipts are useful to communicate the existence of consent decisions, and enable entities to exercise of rights or raise issues and complaint regarding processing activities.

Consent receipts are a relatively newer and under-utilised practice, with no legal requirements existing that refer to the concept (of receipts) or state how they should be used. In addition, the usefulness of receipts as information provided to another entity requires consideration of specific terms and norms particular to the domain or sector. ISO-27560 follows this by providing the flexibility for organisations to choose a suitable schema for their particular domain or use-case. It defines a minimal structure consisting of some fields representing the receipt metadata, but does not have any requirements on the information structuring within the receipt or its correspondence to fields within the record.

A ISO-27560 conformant consent receipt only requires a metadata section providing information about the consent receipt such as its identifier. Deciding on which additional information is to be provided and in what forms and using which structures is left up to implementing entities. In this guide, we presume that the consent receipt is intended for providing a copy of all information within the consent record.

According to ISO-27560, records are generated and maintained by organisations (Controller, Third Party), and are utilised to provide receipts to a Data Subject. In contrast, the Kantara Consent Receipts specification [9], upon which ISO-27560 is based, defines Consent Receipts as being provided by a Data Subject to a Controller.

For practical considerations of this work, we make no presumptions or enact restrictions on the use of records and receipts. Any entity, be it a Controller or a Data Subject, can maintain their own consent record or issue receipts. Though the phrasing of some sections may imply the Controller as the implementing entity, it does not preclude another entity from also implementing ISO-27560.

**Structure.** A Consent Record contains four sections as described below and depicted visually in Fig. 1 (the terms used are based on the implementation of ISO-27560 for GDPR using DPV as described later in the article):

1. **Header Fields:** these provide metadata about the record e.g. its unique identifier and timestamp of creation. These fields also include information on the schema which dictates how the information in the record is structured and which fields are necessary/optional. ISO-27560 permits creation of different schemas to support varying use-cases and information requirements.
2. **Processing Fields:** these provide information about the processing activities e.g. purposes, personal data, storage durations, geographic locations and restrictions, link to privacy notice, rights, and others.
3. **Parties Fields:** these provide information about entities involved in the processing e.g. controllers, third parties, authorities. The party has an identifier which is used to link or associate it with fields in the processing section e.g. to indicate which party is the controller.
4. **Events Fields:** these provide information about *consent events* e.g. consent given, consent withdrawn. Information includes the type of event, time, duration, associated entity, and how it was expressed.

Each section contains fields which describe the information that must be represented along with the form (e.g. timestamp format) and its necessity (e.g. required or optional). Certain fields are expressed as references to other fields (e.g. 'Controller' in 'Processing' section is a reference to an instance or record in 'Parties' section).

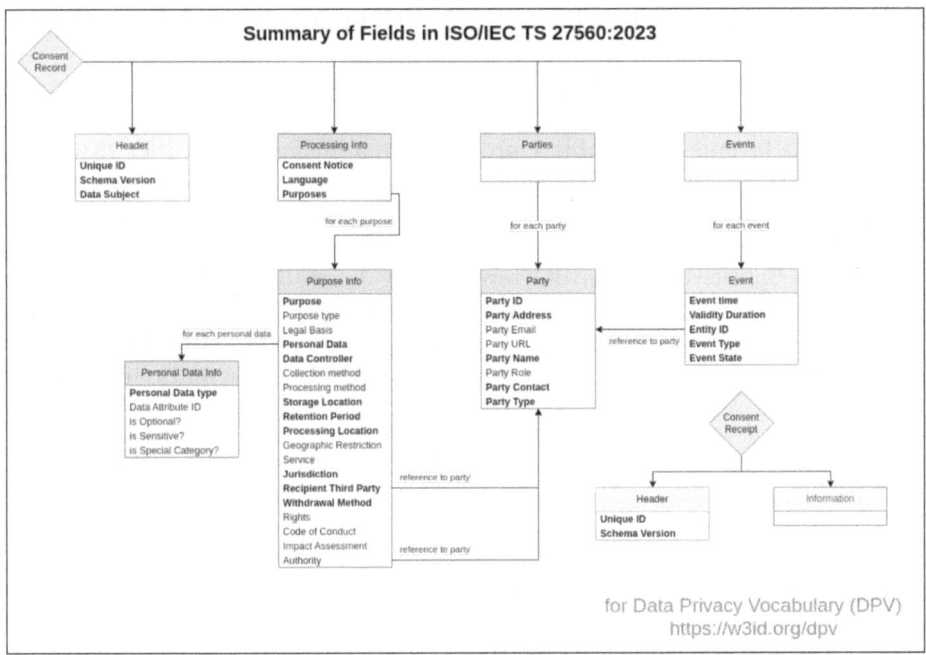

**Fig. 1.** Summary of fields in ISO/IEC TS 27560:2023. The field names have been modified for alignment with DPV concepts. Field names in bold are mandatory.

The Consent Receipt in ISO-27560 contains only two required fields representing a unique identifier for the receipt and the schema version used for the structuring of information. The information and contents are undefined and left to each implementor to specify. A receipt can optionally contain the entirety of the information within a consent record or can also contain multiple consent records or other information not within a particular consent record. Similarly, a receipt can be made to contain only references to information within a record without containing the information itself e.g. providing only the consent record identifier without the contents of the record itself.

Considering the practical application of consent receipts require them to provide information to data subjects, for the implementation described in this document, it is assumed that the consent receipt provides all information contained within a consent record i.e. a receipt is a copy of the record provided to another entity. This is in line with ISO-27560 guidance which states that the receipt may contain the same fields as that of a consent record, and that the mandatory fields in a consent record are also mandatory in a consent receipt. Further, ISO-27560 allows creating different 'schemas' (which we call 'profiles') to indicate changes in requirements and their interpretations, through which we provide profiles for our defined implementations.

# 3   Comparing ISO-27560, ISO-29184, and GDPR

ISO-27560 uses prior terminology established in ISO standards, primarily defined in ISO/IEC 29100:2011 Privacy Framework [4]. To support readers unfamiliar with the ISO terminology, Table 1 provides a mapping between ISO-29100 and GDPR terminology regarding the fundamental concepts associated with personal data processing. Note that this mapping only provides relevant concepts and does not indicate that the concepts are interpreted in the exact same way - for example Sensitive PII in ISO terminology is *similar* to Special Category personal data under GDPR, but they cannot be used interchangeably. Therefore, when applying ISO standards to GDPR, such mappings are indicative of which concepts should be (re-)interpreted with GDPR's definitions and requirements.

**Table 1.** Mapping between ISO/IEC 29100:2011 and EU GDPR terminology

| ISO/IEC 29100:2011 | EU GDPR |
| --- | --- |
| Section 2.4 Consent | Article 4-11 Consent |
| Section 2.9 PII | Article 4-1 Personal Data |
| Section 2.10 PII Controller | Article 4-7 Controller |
| Section 2.11 PII Principal | Article 4-1 Data Subject |
| Section 2.12 PII Processor | Article 4-8 Processor |
| Section 2.23 Processing of PII | Article 4-2 Processing |
| Section 2.26 Sensitive PII | Article 9 Special Categories of Personal Data |
| Section 2.27 Third Party | Article 4-10 Third Party |

In prior work [14], we analysed and compared ISO-29184 requirements for notice and consent with those in GDPR to understand the extent to which ISO-29184 standard can be applied to demonstrate compliance with the requirements of the GDPR. We also explored the possibility of using ISO-29184 certifications under GDPR for consent and notice. In continuation of that work, Table 2 compares ISO-27560 for consent information and ISO-29184 for privacy notice information with the requirements under GDPR to provide a holistic view of how the two standards can be used to address GDPR's requirements. In this, it is important to note that unlike ISO-29184 which is an international standard, ISO-27560 is what ISO terms a *Technical Specification (TS)* which only provides guidance and is intended to obtain feedback to create a (future) international standard.

**Table 2.** Mapping information requirements across ISO/IEC TS 27560:2023, ISO/IEC 29184:2020 and EU GDPR. For GDPR, numbers without prefixes are Articles, and with prefix R are Recitals

| ISO/IEC TS 27560:2023 | ISO/IEC 29184:2020 | EU GDPR |
|---|---|---|
| 3.1 consent | – | 4-11 definition of consent |
| 3.2 consent receipt | Annex B | R42, 7-1 demonstrating consent |
| 3.3 consent record | – | R42, 7-1, 13, 14, 30 recording information related to consent |
| 3.4 consent type | 5.4.3 Informed and freely given consent. 3.1 explicit consent | R32, R43, 6-1a, 9-2a conditions for consent. R42 demonstrating consent. 8 child's consent. 9-2a explicit consent |
| 6.2 recordkeeping for privacy notices and consent | – | R42, 7-1, 13, 14, 30 recording and demonstrating consent |
| 6.2.2.1 presentation of notice | 5.2.2 providing notice, 5.2.3 appropriate expression, 5.2.7 appropriate forms, 5.2.9 accessibility | R32, R42, R43, R58, 7-2 notice for consent |
| 6.2.2.2 timeliness of notice | 5.2.5 appropriate timing | R32, R42, R43, R60, 7-2, 13, 14 notice for providing information and requesting consent. R61, 13-2 14-3 timing of notice. R62 exceptions |
| 6.2.2.3 obtaining consent | 5.2.7 appropriate forms | R42, 7-1 record of consent |
| 6.2.2.4 time and manner of consent | 5.2.6 appropriate locations | R32, R42, R43, 7-2 |
| 6.2.2.5 technical implementation | – | R42, 7-1, 13, 14, 30 maintaining information for demonstrating consent |
| 6.2.2.6 unique reference | 5.2.8 ongoing reference | R42, 7-1 demonstrating consent |
| 6.2.2.7 legal compliance | – | R39, 5 principles, 5 principles. R40, R41, 6 lawfulness and legal basis. R50 further processing. R42, 7-1 record of consent |
| 6.3.4.1 privacy_notice | 3.2 notice | R32, R42, R43, R60, R61, 7-2, 13, 14 notice for providing information |
| 6.3.4.2 language | 5.2.4 multi-lingual notice | R32, R42, R43 conditions for consent |
| 6.3.4.3, 6.3.4.4 purposes | 5.3.2 purpose description, 5.3.3 Presentation of purpose description | 5, 6-1a, 13-1c, 13-3, 14-1c, 14-4, 15-a, 30-1b purpose of processing |
| 6.3.4.6 lawful_basis | 5.3.15 Basis for processing | R40, R41, 6-1a, 7-1a, 9-2a, 13-1c, 13-1d, 14-1c lawfulness and legal basis |
| 6.3.4.7 pii_information | 3.3 element of PII | 4-1, 14-1d, 15-1b, 30-1c personal data |
| 6.3.4.8 pii_controllers | 5.3.4 Identification of the PII controller | 13-1a, 14-1a, 30-1a identity of controller |
| 6.3.4.9 collection_method | 5.3.5 PII collection, 5.3.6 Collection method, 5.3.7 Timing and location of the PII collection | R61, 13-1, 14-1, 14-2f, 15-1g source of personal data |
| 6.3.4.10 processing_method | 5.3.8 Method of use | 4-2, 30-2b processing methods. 13-2f, 14-2g, 15-1h automated decision making and profiling |
| 6.3.4.11 storage_locations | 5.3.9 Geo-location of, and legal jurisdiction over, stored PII | 13-1f, 14-1f, 15-2 storage or processing location |
| 6.3.4.12 retention_period | 5.3.11 Retention period | 13-2a, 14-2a, 15-1d, 30-1f storage duration or time limits |
| 6.3.4.13 processing_locations | 5.3.9 Geo-location of, and legal jurisdiction over, stored PII | 13-1f, 14-1f, 15-2 processing location (including data transfers) |
| 6.3.4.14 geo-graphic_restrictions | 5.3.9 Geo-location of, and legal jurisdiction over, stored PII | 13-1f, 14-1f, 15-2, 30-1e, 30-2c, 44, 45, 46, 47, 48, 49-1a geographic condition (e.g. third country) |

continued

**Table 2.** continued

| | | |
|---|---|---|
| 6.3.4.16 jurisdiction | 5.3.9 Geo-location of, and legal jurisdiction over, stored PII | 13-1f, 14-1f, 15-2, 30-1e, 30-2c, 44, 45, 46, 47, 48, 49-1a geographic condition (e.g. third country) |
| 6.3.4.17 recipient_third_parties | 5.3.10 Third-party transfer | 4-9, 4-10, 13-1e, 14-1e, 15-1c, 19, 30-1d recipients |
| 6.3.4.18 withdrawal_method | 5.3.12 Participation of PII principal | R42, 7-3, 13-2c, 14-2d withdrawing consent |
| 6.3.4.19 privacy_rights | 5.3.12 Participation of PII principal | 13-2b, 13-2c, 14-2c, 14-2d, 15-1e, 16, 17, 18, 20, 21, 22 rights of data subject |
| 6.3.4.20 codes_of_conduct | – | 24-3, 32-3, 35-8, 40 codes of conduct, 42 certification |
| 6.3.4.21 impact_assessment | 5.3.16 Risks | R75, R84 risks and risk evaluation. R90, R91, R92, R93, 35 Data Protection Impact Assessments (DPIA) |
| 6.3.4.22 authority_party | 5.3.13 Inquiry and complaint | 13-2d, 14-2e, 15-1f complaint to authority. 36-1 consult with authority for impact assessment. 51 supervisory authority, 56 lead authority. |
| 6.3.5.1 pii_type | 3.3 element of PII | 4-1, 14-1d, 15-1b, 30-1c personal data types and categories |
| 6.3.5.2 pii_attribute_id | 3.3 element of PII | – |
| 6.3.5.3 pii_optional | 5.4.6 Separate consent to necessary and optional elements of PII | R90, R91, 5, 13-2e, 35 optionality or necessity of personal data |
| 6.3.5.4 sensitive_pii_category | – | R51 protecting sensitive data |
| 6.3.5.5 special_pii_category | – | R51, R53, R71, 6, 9, 22-4, 30-1c, 35 special categories of personal data |
| 6.3.6.6 party_name | – | 13-1a, 14-1a, 30-1a, 30-2a |
| 6.3.6.7 party_role | – | 4-1, 4-7, 4-7, 4-8, 4-9, 4-10, 13-1a, 13-1e, 14-1a, 14-1e, 26-1, 28, 30-1a, 30-1d, 30-2a, 37 |
| 6.3.6.8 party_contact | – | 13-1a, 13-1b, 14-1a, 14-1b, 26-1 |
| 6.3.6.9 party_type | – | 4-1, 4-7, 4-7, 4-8, 4-9, 4-10, 13-1e, 14-1e, 15-1c, 19 |
| 6.3.7.1 event_time | 5.4.8 Timeliness | R42, 7-1, 13, 14, 30 maintaining information for demonstrating consent |
| 6.3.7.2 validity_duration | 5.4.7 Frequency | 25 Data Protection by Design and by Default |
| 6.3.7.3 entity_id | – | R42, 4-11, 6-1a, 7-3, 8-1, 8-2, A13, A14 |
| 6.3.7.4 event_type | 5.4.3 Informed and freely given consent | 4-11 (regular) consent. 9-2a explicit consent. R32, 7-1 given consent. R32, 7-2 request for consent |
| 6.3.7.6 event_state | 5.5.2 Renewing notice, 5.5 Change of conditions, 5.5.3 Renewing consent | 4-11, 6-1a, 9-2a given consent. R42, 7-3 withdrawn consent. |
| 6.4.3 consent management | – | R32, R42, R43, R60, R61, 7-2, 12, 13, 14 information about given consent and applicable rights, R42, 7-3 withdrawing consent |
| 6.4.4 PII principal participation | 5.3.12 Participation of PII principal, 5.3.14 Information about accessing the choices made for consent | R32, R42, R43, R60, R61, 7-2, 12, 13, 14 information about given consent and applicable rights, R42, 7-3 withdrawing consent |
| 6.4.6 receipt content | 5.3.14 Information about accessing the choices made for consent | R32, R42, R43, R60, R61, 7-2, 12, 13, 14 information about given consent and applicable rights |

continued

**Table 2.** continued

| Annex B consent lifecycle | 5.5.2 Renewing notice, 5.5 Change of conditions, 5.5.3 Renewing consent | 4-11, 6-1a, 9-2a given consent. R42, 7-3 withdrawn consent. |
|---|---|---|
| Annex E security of consent records and receipts | – | R75, R76, R77, R78, R83, 24, 25, 30, 32, 44 |
| Annex F signals communicating PII Principal's preferences and decisions | – | R32, 7-2, 21-5 |

# 4    Consent Records and Receipts Using DPV

ISO-27560 only defines the information fields and does not prescribe how they should be technically represented in practice. To implement ISO-27560, the information therefore must be represented in a format such as JSON which is widely supported and easy to use. However, the use of JSON requires a strict agreement on how the information should be structured and how it should be interpreted. The JSON-LD (JSON for Linked Data) format enables use of JSON with linked data so that the ontologies and vocabularies defined using W3C standards can be exchanged as JSON data. For this reason, ISO-27560 Annex C provides examples of consent records and receipts for both JSON and JSON-LD. Implementing ISO-27560 in a machine-readable manner using JSON-LD requires agreement on the schema or ontology to represent the fields. The Annex C JSON-LD example uses the Data Privacy Vocabulary[1] (DPV) [13,15] which is maintained by the W3C Data Privacy Vocabularies and Controls Community Group[2] (DPVCG).

DPV is a state of the art resource that provides the necessary ontology to represent concepts such as purpose, processing operations, personal data, legal roles, as well as a rich and extensive taxonomy expanding on each of these concepts to enable representing of practical use-cases. For example, using DPV, it is possible to exchange ISO-27560 records and receipts in JSON-LD which specify the *Purpose* is *Marketing* in an interoperable manner. In addition, DPV also features extensions through which different jurisdiction specific concepts can be represented, and for which it provides extensions for EU regulations such as the GDPR, DGA, and the upcoming AI Act. This enables flexibility of expression general requirements such as the *legal basis* should be *consent*, as well as specific requirements such as *explicit consent* as per GDPR Art.9-1a.

DPV was initiated as part of the SPECIAL H2020 project and has been developed for over 6 years with a multi-disciplinary community made up of computer scientists, legal experts, sociologists, data protection officers, industry stakeholders, and authorities. It has been actively used in several projects at national and international (e.g. Horizon Europe) levels, is being used by the industry, and has been acknowledged by within standards (including ISO-27560) [13]. As such, it represents the best resource currently available for representing consent records

---

[1] https://w3id.org/dpv.
[2] https://www.w3.org/groups/cg/dpvcg/.

and receipts as well as other legally relevant information in a machine-readable form that is based on open (free and non-proprietary) interoperable standards.

To support the implementation of use of DPV in implementing consent records and receipts in conformance with ISO-27560 and the GDPR, we have developed a technical specification which can be accessed online at https://w3id.org/dpv/guides/consent-27560. The specification describes how the required fields in ISO-27560 and GDPR are represented using DPV (summarised below) and provides illustrative examples for each (see online). Consent records are represented as instances of the concept `dpv:ConsentRecord` and receipts are represented as instances of the concept `dpv:ConsentReceipt`. For reviewers convenience: complete examples of a consent record and a consent receipt are provided in the annexes at the end of this article.

**Profiles:** To support implementing ISO-27560 as well as its use to comply with GDPR, DPVCG defines 4 schemas or profiles defined under the namespace https://w3id.org/dpv/schema/dpv-27560# (prefixed as `dpv-27560:`).

1. dpv-27560:record: Consent Records conforming with ISO-27560.
2. dpv-27560:record-eu-gdpr Consent Records conforming with ISO-27560 and containing information as required by EU GDPR.
3. dpv-27560:receipt-record Consent Receipts conforming with ISO-27560 and providing information from consent record(s).
4. dpv-27560:receipt-eu-gdpr Consent Receipts conforming with ISO-27560 and providing information from consent record(s) as required by EU GDPR.

**Metadata Fields:** (See Table 3) to describe the generic metadata fields associated with records and receipts, we utilise the DCMI Metadata Terms standard[3] (prefixed as `dct:`). A consent record or receipt indicates use of the DPV profiles by using `dct:conformsTo` with one of the profiles described above.

**Table 3.** DPV concepts for ISO/IEC 27560:2023 Metadata fields

| Field | Cardinality | DPV Concept | DPV Property |
|---|---|---|---|
| Schema Version | 1 | N/A | extttdct:conformsTo |
| Record Identifier | 1..* | N/A | `dpv:hasIdentifier` |
| Data Subject | 1 | `dpv:DataSubject` | `dpv:hasDataSubject` |

**Processing Fields:** (See Table 4) ISO-27560 contains 22 fields related to processing activities, and 5 additional fields regarding personal data involved in processing. The structuring of these fields within ISO-27560 is of the form where

---

[3] https://dublincore.org/specifications/dublin-core/dcmi-terms/.

the 'PII Processing' section contains an array of 'purposes' where each 'purpose' is expressed with its own fields regarding legal basis, collection method, storage locations, and so on. Within the DPV implementation, this is replaced with `dpv:Process` where each 'process' represents a distinct processing activity with its own fields e.g. purposes, personal data, recipients. Thus a consent record and receipt may cover multiple processes (and purposes) which permits an unambiguous and exact representation e.g. which purpose, implemented by which entity, with what data, recipient, etc.

**Table 4.** DPV concepts for ISO/IEC 27560:2023 Processing fields

| Field | Cardinality | DPV Concept | DPV Property |
|---|---|---|---|
| Process | 1..* | dpv:Process | dpv:hasProcess |
| Purpose | 1..* | dpv:Purpose | dpv:hasPurpose |
| Personal Data | 1..* | dpv: | dpv:hasPersonalData |
| Personal Data Type | 1..* | dpv:PersonalData taxonomy | dpv:hasPersonalData or dct:type |
| Personal Data Identifier | 0..* | N/A | dct:identifier |
| Personal Data Necessity | 0..* | dpv:Necessity | dpv:hasNecessity |
| Sensitive/Special Category | 0..* | dpv:SensitivePersonalData, dpv:SpecialCategoryPersonalData | dpv:hasPersonalData or dct:type |
| Processing Operations | 0..* | dpv:Processing | dpv:hasProcessing |
| Data Source | 0..* | dpv:DataSource | dpv:hasDataSource |
| Storage Condition | 1..* | dpv:StorageCondition, dpv:StorageLocation, dpv:StorageDuration, dpv:StorageDeletion | dpv:hasStorageCondition |
| Processing Condition | 0..* | dpv:ProcessingCondition, dpv:ProcessingLocation, dpv:ProcessingDuration | dpv:hasProcessingCondition |
| Geographic Restriction | 0..* | dpv:Rule | dpv:hasRule |
| Data Controller | 1..* | dpv:DataController | dpv:hasDataController |
| Legal Basis | 0..* | dpv:LegalBasis | dpv:hasLegalBasis |
| Recipients | 1..* | dpv:Recipient | dpv:hasRecipient |
| Consent Change &amp; Withdrawal | 1..* | dpv:InvolvementControl, dpv:WithdrawingFromActivity | dpv:hasInvolvementControl |
| Jurisdiction | 1..* | dpv:Jurisdiction | dpv:hasJurisdiction |
| Rights | 1..* | dpv:DataSubjectRight | dpv:hasRight |
| Services | 0..* | dpv:Service | dpv:hasService |
| Code of Conduct | 0..* | dpv:CodeOfConduct | dpv:hasOrganisationalMeasure |
| Impact Assessment | 0..* | dpv:ImpactAssessment | dpv:hasAssessment |
| Notice | 1..* | dpv:Notice | dpv:hasNotice |
| Notice Language | 1..* | N/A | dct:language |

**Entity Fields:** (See Table 4) DPV uses the term *Entity* for what ISO-27560 refers to as *Party*. Entities are expressed using instances of `dpv:Entity` and associated using `dpv:hasEntity`. DPV also distinguishes between *Entities* and *Legal Entities* - and their representatives or agents, through which it can be accurately represented whether a party in a consent record acted on their own or it was someone acting on their behalf. This is of relevance for implementations such as consent for children which involves parents or guardians, or even data intermediaries under DGA which can act to support individuals in consent decision making.

**Table 5.** DPV concepts for ISO/IEC 27560:2023 Party fields

| Field | Cardinality | DPV Concept | DPV Property |
|---|---|---|---|
| Name | 1..* | N/A | `dpv:hasName` |
| Identifier | 1..* | N/A | `dpv:hasIdentifier` |
| Role | 1..* | `dpv:DataController,` `dpv:DataProcessor,` `dpv:ThirdParty,` `dpv:Authority,` `dpv:DataSubject` | `dpv:hasEntity,` `dpv:hasDataController,` `dpv:hasDataProcessor,` `dpv:hasThirdParty,` `dpv:hasAuthority,` `dpv:hasDataSubject` |
| Contact | 1..* | `schema:ContactPoint` | `schema:contactPoint` |
| Postal Address | 1..* | `schema:PostalAddress` | `schema:contactPoint` |
| Email | 0..* | N/A | `schema:email` |
| Phone | 0..* | N/A | `schema:telephone` |
| URL | 0..* | N/A | `schema:url` |

**Consent Event Fields:** (See Table 5) These fields are used to indicate the type of consent (e.g. *Implicit, Expressed, Explicit*) as expressed by the data subject. In DPV, `dpv:ConsentType` represents consent types to be used as a legal basis and has the following different types: *Informed* with specialisations for *Implied* when implied or given by an indirect action (e.g. merely browsing a website), *Expressed* for direct expressed action (e.g. a checkbox), and *ExplicitlyExpressed* for direct action concerning solely the consent in context. To indicate consent types as defined in GDPR, the DPV's GDPR extension is used e.g. `eu-gdpr:A6-1-a` for expressed consent and `eu-gdpr:A9-2-a` for explicit consent.

In addition to the type of consent, these fields also enable expressing the status of consent, such as whether it has been requested, given, refused, expired, terminated, invalidated, or re-affirmed. Each event can contain metadata to indicate when it took place (e.g. date when consent was given), how it was expressed (e.g. in the account dashboard), its duration (e.g. validity of given consent), and by whom (e.g. the data subject).

**Table 6.** DPV concepts for ISO/IEC 27560:2023 Event fields

| Field | Cardinality | DPV Concept | DPV Property |
|---|---|---|---|
| Consent Type | 1..* | `dpv:Consent` taxonomy | `dpv:hasLegalBasis` |
| Consent State | 1..* | `dpv:ConsentStatus` taxonomy | `dpv:hasConsentStatus` |
| Event Time | 1..* | N/A | `dpv:isIndicatedAtTime` |
| Event Duration | 1..* | `dpv:Duration` | `dpv:hasDuration` |
| Expression by Entity | 1..* | `dpv:Entity` | `dpv:isIndicatedBy` |
| Expression Method | 0..* | N/A | `dpv:hasIndicationMethod` |

**Consent Receipts:** (See Table 7) ISO-27560 only defines the schema version and receipt identifier fields for consent receipts. For other fields, it recommends using the same fields as that of a consent record. In its guidance, it states that the mandatory fields in consent records should also be mandatory in receipts. Based on this, we only define the additional fields for consent receipts and suggest reusing the consent record fields with their necessity/optionality requirements. Therefore, a consent receipt only has three mandatory fields with the rest of the information being present as instances of consent record(s).

**Table 7.** DPV concepts for ISO/IEC 27560:2023 Receipt Metadata fields

| Field | Cardinality | DPV Concept | DPV Property |
|---|---|---|---|
| Schema Version | 1 | N/A | `dct:conformsTo` |
| Receipt Identifier | 1..* | N/A | `dpv:hasIdentifier` |
| Consent Record | 1..* | `dpv:ConsentRecord` | `dpv:hasRecordOfActivity` |

## 5 Supporting GDPR and DGA

**Using ISO-27560 and ISO-29184 Within the EU Legal Framework:** ISO-27560 and ISO-29184 are developed and governed by the International Standards Organisation (ISO), and are not specific to EU's regulations and terminology. To support their use in the legal frameworks, they need to be approved as 'Euronorm' (EN) through an EU standardisation body such as CEN, CENELEC, or ESO. At the moment, ISO-29184 has already been approved as EN, and we are working on a proposal with the Irish and Swedish national bodies to recommend the adoption of ISO-27560 as EN. Further, we have also submitted a proposal to the relevant ISO committees to make ISO-27560 standard freely accessible as its guidance is valuable for responsible innovation.

Having these standards as EN provides a strong framework for their utilisation in regulations, such as for notice and consent under GDPR. However, merely adopting the standards on an 'as-is' basis will not be sufficient. For example, the terminology in 29184 and GDPR has crucial differences which must be identified and appropriate guidance developed to enable using ISO-29184 with GDPR [14]. Similarly, to address current issues regarding consent [10,11] and further studies are required to assess the extent of these standards in solving existing issues and what additional measures need to be adopted beyond conformance with the standards.

**Demonstrating Consent Under GDPR:** GDPR Article 7-1 creates an obligation for data controllers to maintain consent information and to keep it up to date with the goal of demonstrating where consent was given, refused, or withdrawn. ISO-27560 provides a standard for a common technical structure to support implementing this obligation. In addition to this, GDPR Article 13 and Article 14, amongst others, also require record keeping for what information was provided to individuals in order to implement informed consent. ISO-29184 provides a standard for describing privacy notices, and together with ISO-27560 enables maintaining records of what information was provided and the resulting consent decisions. Based on the analysis provided in this article that demonstrates applicability of ISO-27560 and ISO-29184 to GDPR, we recommend authorities to suggest using these standards to support GDPR compliance.

**Receipts to Support Rights Under GDPR:** ISO-27560 contains fields for acknowledging which rights exist, and with DPV we can express how/where to exercise them and what information will be required (e.g. identity verification). Further, consent decisions (e.g. given, withdrawn) are themselves also personal data about the data subject, and therefore subject to rights such as Article 20 data portability. This can be a way to enable the use of receipts under GDPR even where it is not explicitly defined as a concept by considering consent information as *personal data*. Considering consent information as personal data makes it subject to the right to data portability under Article 20 which requires providing information "in a structured, commonly used and machine-readable format". Further, Article 20 also allows "the right to transmit those data to another controller", which can be utilised to transfer consent decisions from one controller to another - a crucial mechanism for the implementation of data reuse and altruism under DGA.

**Common Consent Form Under DGA:** Article 25 of the DGA requires the Commission to produce a common consent form that will provide information in both human- and machine-readable forms. ISO-27560 with ISO-29184, based on the analysis in this article demonstrating their usefulness to meet GDPR requirements, should be used to define what information should be present in these forms. ISO-29184 as the standard for privacy notices provides the human-oriented representation of information in the consent form, and ISO-27560 and the DPV implementation provide the machine-readable representation. The advantage of using these standards is that the resulting solution would be useful not only in EU but globally due to the global scope of ISO. The advantage of using DPV here is in providing common semantics based on W3C standards that support extensions for specific jurisdictions (like EU with GDPR and DGA) and its extensive taxonomy supporting practical use-cases which promote interoperability. **Through direct meetings, we have presented this work to the EU Commission's Unit G.1 which looks after GDPR and DGA implementations.**

**Data Intermediaries Under DGA:** We are also working on further implementations to support DGA by developing specific technical specifications that define how data intermediaries should maintain consent records and issue receipts, and support them in their duties by providing a way to express data reuse requests in a machine-readable form that can be matched with the consent to ensure the purposes are compatible in accordance with the GDPR. This will be based on existing work [1] that utilises the W3C Open Digital Rights Language (ODRL) standard [3] for representing policies and agreements, and using it in combination with DPV to create DGA specific *offers* for data subjects and data intermediaries to indicate which data is available for reuse and under what conditions, requests for data users to indicate what data they are looking for, and *agreements* to represent the conditions under which data reuse has been approved. We have already demonstrated the feasibility of using ODRL and DPV for such an approach in a manner that improves both technical and organisational processes for the use-case of sharing genomic health datasets [12].

**Data Reuse and Altruism Under DGA:** To support the DGA's goals of reusing data for altruism, we are working on creating a taxonomy of altruistic purposes within DPV and developing a framework to express them in a manner that is compatible with GDPR's requirements for consent and information keeping based on ISO-27560. We are also working on novel approaches such as assessing the compatibility of ISO-27560 defined consent records with information required in a Data Protection Impact Assessment (DPIA), through which we aim to enable data subjects or data intermediaries to conduct their own DPIAs based on a common registry of risks and mitigations provided through the DPV. Through this we aim to establish responsible practices while promoting data reuse and altruism.

## 6    Implementation Considerations and Future Work

### 6.1    Trust and Security

Security considerations are extremely important in the implementation of consent records and receipts, with ISO-27560 Annex E providing guidance for implementations. Consent records are intended to be maintained internally by an entity, and require measures to ensure they maintain their consistency and correctness, and are not tampered with. This includes best practices for information management such as using cryptographic hashes to ensure information has not changed, or using access control to ensure only authorised modifications are permitted. Current internationals standards such as W3C Decentralized Identifiers[4] (DID) and W3C Verifiable Credentials[5] (VC) allow for implementations compatible with the implementation of ISO-27560 using DPV as all are based on interoperable semantic web standards.

For consent receipts to be utilised in a verifiable and trustworthy manner, the information provided within the receipt may require cryptographic measures

---

[4] https://www.w3.org/TR/vc-data-model/.
[5] https://www.w3.org/TR/did-core/.

to provide assurance to prove its immutability and non-repudiation. Further, receipts are intended to be information provided or exchanged between different entities, which may necessitate a mechanism to demonstrably verify the provenance (e.g. a receipt was provided by A to B) and its immutability (e.g. receipt contained X exactly). Cryptography techniques such as digital signatures and certificates can support such applications based on their current utilisation in internet-enabled applications and documentations. Prior work [7] and projects[6] have explored such considerations, but effective implementation requires consensus amongst stakeholders to create an interoperable ecosystem.

Given the role of consent records and receipts in demonstrating consent decisions, they may end up with potentially sensitive information. ISO-27560 recommends not putting such information directly in records and receipts, and if necessary then implementations should utilise techniques such as information masking or pseudonymisation to avoid directly exposing sensitive information. - though this has to be balanced with the purpose of receipts in providing data subjects with information about their consent.

## 6.2    Using Records and Receipts with eIDAS and EUDI Wallet

Following the launch of projects for using European Digital Identity wallet (EUDI) wallet[7] for travel, health, banking, education and other sectors, CEN TC224 WG20[8], which is the EU standardisation body's technical committee for personal identification, has initiated a new standards project to provide guidance on when personal data (attributes) are shared from the wallet in compliance with eIDAS and its proposed revision.

In this, ISO-27560 and ISO-29184 can be used to create an interoperable and standards based mechanism to structure information and ensure the mandatory fields needed to comply with GDPR are present. Further, the use of these standards also enables a consistent approach for creating common privacy dashboards that can work across EU. Such privacy dashboards would allow a wallet holder to have an overview of all their consent transactions, including any pending requests as well as provide a centralised mechanism for controlling their rights and withdrawing consent by using the eIDAS and eID mechanisms to establish identity and proof of past engagement.

ISO-27560 and ISO-29184 are also crucial as being the only standards regarding consent records and receipts, and privacy notices respectively. Using the analysis and implementations described in this article, a ISO-27560 solution that is also conformant with the GDPR can be used to store consent records and receipts in wallets, which enables data subjects to have a copy of their decision and agreement to process personal data.

---

[6] NGI funded Privacy as Expected: Consent Gateway project D2 Final Technical Deliverable https://doi.org/10.5281/zenodo.5086238.

[7] https://digital-strategy.ec.europa.eu/en/news/eu-digital-identity-4-projects-launched-test-eudi-wallet.

[8] https://www.cencenelec.eu/areas-of-work/cen-sectors/digital-society-cen/information-and-identification-systems/.

Having this information made available to the data subject in a machine-readable format further enables its use in innovative applications that promote reuse of data while ensuring adequate adherence to the EU's values and regulations. For example, by looking at past consent records or receipts, preferences can be identified for how the individual makes decisions and these can be used to create a template or pattern that will make future consent decisions more efficient and simpler for the individual. ISO-27560 Annex F provides guidance on how such preferences used as 'privacy signals' can be represented within consent records and receipts.

Another powerful paradigm is also made possible when combining ISO-27560 with eID, eIDAS, and EUDI - where the data subject initiates the consent process by providing a specific consent to use or reuse their personal data, for example to access a particular service. In this scenario, the data subject decides the extent and limit of what their consent will cover, provides their consent to the service provider, and maintains a consent record within their wallet with a signed receipt provided to the service provider as proof of consent.

### 6.3 Standard for PII Processing Record Information

Even though ISO-27560 only focuses on consent records and receipts, its fields were developed with the intention of a future expansion in a separate standard to cover other legal bases, such as the 7 other legal basis in GDPR Article 6. To continue in this direction, we have initiated a 'new standard' proposal in ISO regarding 'PII processing record information'.

To support this activity, we are currently identifying the specific requirements for record keeping for each legal basis and creating the necessary specifications using DPV. This builds on prior work providing a machine-readable Records of Processing Activities (ROPA) required under GDPR Article 30, and which consolidates the guidelines from all 30 EU/EEA member states.

### 6.4 Technical Considerations in Managing Records and Receipts

We can use the Data Catalog Vocabulary[9] (DCAT), a W3C standard, to represent the records as datasets and receipts as a catalogue of records. By doing so, the metadata fields provided by DCAT can be readily used to represent information that supports in maintenance and exchange of consent records and receipts, including using existing infrastructure to manage them. DCAT is a widely used standard that supports implementing (open) data portals and has tooling for discovery and management of information. The EU has developed the DCAT Application Profile[10] (DCAT-AP) which extends DCAT to support the EU Open Data Portals[11].

---

[9] DCAT - Version 3 https://www.w3.org/TR/vocab-dcat-3/.

[10] DCAT Application profile for data portals in Europe (DCAT-AP) https://op.europa.eu/en/web/eu-vocabularies/dcat-ap.

[11] https://data.europa.eu/.

Through these records and receipts can be readily communicated as interoperable datasets between relevant entities - for example controller to data subject, or between controllers and third parties. This is a crucial technical enabler for the principle of increasing data value through utilisation within the Data Governance Act and Data Spaces. In particular, the use of DCAT(-AP) also supports the addition of further policies and measures to support the implementation of data intermediaries which will be required to maintain consent records under the obligations of the DGA.

### 6.5   IEEE P7012 Machine-Readable Privacy Terms

In addition to the above, we are also working with the IEEE P7012 group to develop a standard for machine-readable privacy terms which uses ISO-27560 and ISO-29184 with DPV to define the conditions under which the individual allows use or reuse of their personal data. The use of this standard will provide an efficient and optimal mechanism for data subjects to signal their consent or initiate an agreement with a service provider.

## 7   Conclusion

This article provided a thorough analysis of how ISO/IEC TS 27560:2023 and ISO/IEC 29184:2020 can be used to create consent records and receipts in a machine-readable format that support GDPR requirements and enable the reuse of data under the DGA. Based on this analysis, we provide a concrete argument for why these two standards should be adopted and recommended by GDPR stakeholders. We also described the ongoing efforts of the W3C Data Privacy Vocabularies and Controls Community Group (DPVCG) in creating a technical specification to support implementing ISO-27560 by using its Data Privacy Vocabulary (DPV). Our work is a significant contribution to the ongoing efforts of implementing the DGA where the Commission is required to develop a common consent form that is both human- and machine-readable. We also discussed how this work can be utilised in practice, where reported on our ongoing efforts to adopt the standard within the EU's legal framework, further develop specific implementations to support the needs of DGA, and how this work compliments the ongoing developments of eID, eIDAS2, and EUDI implementations.

**Acknowledgements.** Jan Lindquist, through the Swedish National Standards Body, was a contributor and the co-editor of ISO/IEC TS 27560:2023. Harshvardhan J. Pandit, through the Irish National Standards Body, was a contributor to the ISO/IEC TS 27560:2023, and is the chair of the W3C Data Privacy Vocabularies and Controls Community Group.

This research was conducted with the financial support of Science Foundation Ireland at ADAPT, the SFI Research Centre for AI-Driven Digital Content Technology at Dublin City University Grant#13/RC/2106_P2. For the purpose of Open Access, the author has applied a CC BY public copyright licence to any Author Accepted Manuscript version arising from this submission.

# A  Example of Consent Record with both Required and Optional Fields

```
1   {
2       "@id": "https://example.com/a6f58318-72e6-46a2-bfd7-f36d795e30cd",
3       "@type": "dpv:ConsentRecord",
4       "dct:identifier": "a6f58318-72e6-46a2-bfd7-f36d795e30cd",
5       "dct:conformsTo": "https://w3id.org/dpv/schema/dpv-27560#record",
6       "dpv:hasDataSubject": {
7           "@id": "0760c9ba",
8           "type": "dpv:Consumer"
9       },
10      "dpv:hasDataController": "ex:Acme",
11      "dpv:hasDataProcessor": "ex:Beta",
12      "dpv:hasJurisdiction": ["loc:IE"],
13      "dpv:hasApplicableLaw": "eu-gdpr:GDPR",
14      "dpv:hasLegalBasis": "eu-gdpr:A6-1-a",
15      "dpv:hasProcess": {
16          "@type": "dpv:Process",
17          "dpv:hasService": "Register for Event X",
18          "dpv:hasRecipient": ["ex:Acme", "ex:Beta"],
19          "dpv:hasPurpose": "dpv:PaymentManagement",
20          "dpv:hasPersonalData": {
21              "@type": "pd:EmailAddress",
22              "rdf:value": "hello@example.com",
23              "dpv:hasNecessity": "dpv:Optional",
24              "dpv:hasDataSource": "dpv:DataSubject",
25          },
26          "dpv:hasStorageCondition": [{
27              "@type": "dpv:StorageLocation",
28              "dpv:hasLocation": ["loc:IE", "loc:FR", "loc:DE"],
29          }, {
30              "@type": "dpv:StorageDuration",
31              "dpv:hasDuration": "P6M",
32          }, {
33              "@type": "dpv:StorageDeletion",
34              "dpv:hasDuration": "P1M"
35          }]
36      },
37      "dpv:hasProcess": {
38          "@type": "dpv:Process",
39          "dpv:hasService": "Register for Event X",
40          "dpv:hasRecipient": ["ex:Acme", "dpv:DataSubject"],
41          "dpv:hasPurpose": "dpv:IdentityVerification",
42          "dpv:hasPersonalData": {
43              "@type": "pd:OfficialID",
44              "dct:identifier": "XJ189019D",
45              "dpv:hasNecessity": "dpv:Required",
46              "dpv:hasDataSource": "ex:Acme",
47          },
48          "dpv:hasStorageCondition": [{
49              "@type": "dpv:StorageLocation",
50              "dpv:hasLocation": "dpv:WithinDevice",
```

```
51          }, {
52              "@type": "dpv:StorageDuration",
53              "dpv:hasDuration": {
54                  "@type": "dpv:UntilEventDuration",
55                  "rdf:value": "Account Closure"
56              }]
57          },
58          "dpv:hasNotice": {
59              "@id":
            ↪   "https://example.com/notices/a6f58318-72e6-46a2-bfd7-f36d795e30cd",
60              "@type": "dpv:ConsentNotice",
61              "dct:date": "2024-01-01",
62              "dct:language": "EN",
63              "dct:coverage": "2024-01-01/P12M"
64          }
65          "dpv:hasImpactAssessment": {
66              "@type": "dpv:DPIA",
67              "schema:url": "https://example.com/DPIA"
68          }
69          "dpv:hasInvolvementControl": {
70              "@type": ["dpv:ProvidingPermission", "dpv:WithdrawingPermission"],
71              "dpv:isExercisedAt": "https://example.com/manage-consent"
72          },
73          "dpv:hasRight": [{
74                  "@type": ["dpv:DataSubjectRight", "eu-gdpr:A7-3"],
75                  "dct:title": "Right to Withdraw Consent",
76                  "dpv:isExercisedAt": "https://example.com/rights",
77          },
78          "dpv:hasConsentStatus": [{
79              "@type": ["dpv:ConsentGiven", "dpv:ExpressedConsent"],
80              "dpv:isIndicatedBy": "dpv:DataSubject",
81              "dpv:hasIndicationMethod": "Interaction in App",
82              "dpv:isIndicatedAtTime": "2024-01-01"
83          }, {
84              "@type": "dpv:ConsentWithdrawn",
85              "dpv:isIndicatedBy": "dpv:DataSubject",
86              "dpv:hasIndicationMethod": "Interaction in App",
87              "dpv:isIndicatedAtTime": "2024-04-20"
88          }]
89  }
```

## B   Example of Consent Receipt with Required Fields from Consent Record

```
1  {
2      "@id": "https://example.com/receipt-asdmj1oasd",
3      "@type": "dpv:ConsentRereceipt",
4      "dct:identifier": "receipt-asdmj1oasd",
5      "dct:conformsTo": "https://w3id.org/dpv/schema/dpv-27560#receipt",
6      "dct:created": "2024-01-31",
```

```
 7        "dct:publisher": "ex:Acme",
 8        "schema:recipient": "dpv:DataSubject",
 9        "dpv:hasRecordOfActivity": {
10            "@id": "htttps://example.com/a6f58318-72e6-46a2-bfd7-f36d795e30cd",
11            "@type": "dpv:ConsentRecord",
12            "dct:identifier": "a6f58318-72e6-46a2-bfd7-f36d795e30cd",
13            "dct:conformsTo": "https://w3id.org/dpv/schema/dpv-27560#record",
14            "dpv:hasDataSubject": {
15                "@id": "0760c9ba",
16                "type": "dpv:Consumer"
17            },
18            "dpv:hasDataController": "ex:Acme",
19            "dpv:hasDataProcessor": "ex:Beta",
20            "dpv:hasJurisdiction": ["loc:IE"],
21            "dpv:hasApplicableLaw": "eu-gdpr:GDPR",
22            "dpv:hasProcess": {
23                "@type": "dpv:Process",
24                "dpv:hasRecipient": ["ex:Acme", "ex:Beta"],
25                "dpv:hasPurpose": "dpv:PaymentManagement",
26                "dpv:hasPersonalData": "pd:EmailAddress",
27                "dpv:hasStorageCondition": [{
28                    "@type": "dpv:StorageLocation",
29                    "dpv:hasLocation": ["loc:IE", "loc:FR", "loc:DE"]
30                }, {
31                    "@type": "dpv:StorageDuration",
32                    "dpv:hasDuration": "P6M"
33                }, {
34                    "@type": "dpv:StorageDeletion",
35                    "dpv:hasDuration": "P1M"
36                }]
37            },
38            "dpv:hasProcess": {
39                "@type": "dpv:Process",
40                "dpv:hasRecipient": ["ex:Acme", "dpv:DataSubject"],
41                "dpv:hasPurpose": "dpv:IdentityVerification",
42                "dpv:hasPersonalData": "pd:OfficialID",
43                "dpv:hasStorageCondition": [{
44                    "@type": "dpv:StorageLocation",
45                    "dpv:hasLocation": "dpv:WithinDevice"
46                }, {
47                    "@type": "dpv:StorageDuration",
48                    "dpv:hasDuration": {
49                        "@type": "dpv:UntilEventDuration",
50                        "rdf:value": "Account Closure"
51                    }
52                }]
53            },
54            "dpv:hasInvolvementControl": {
55              "@type": ["dpv:ProvidingPermission", "dpv:WithdrawingPermission"],
56              "dpv:isExercisedAt": "https://example.com/manage-consent"
57            },
58            "dpv:hasRight": {
59                "@type": ["dpv:DataSubjectRight", "eu-gdpr:A7-3"],
60                "dct:title": "Right to Withdraw Consent",
```

```
61              "dpv:isExercisedAt": "https://example.com/rights"
62          },
63          "dpv:hasNotice": {
64              "@id":
              ↪ "https://example.com/notices/a6f58318-72e6-46a2-bfd7-f36d795e30cd",
65              "@type": "dpv:ConsentNotice",
66              "dct:date": "2024-01-01",
67              "dct:language": "EN",
68              "dct:coverage": "2024-01-01/P12M"
69          },
70          "dpv:hasConsentStatus": [{
71              "@type": ["dpv:ConsentGiven", "dpv:ExpressedConsent"],
72              "dpv:isIndicatedBy": "dpv:DataSubject",
73              "dpv:hasIndicationMethod": "Interaction in App",
74              "dpv:isIndicatedAtTime": "2024-01-01"
75          }, {
76              "@type": "dpv:ConsentWithdrawn",
77              "dpv:isIndicatedBy": "dpv:DataSubject",
78              "dpv:hasIndicationMethod": "Interaction in App",
79              "dpv:isIndicatedAtTime": "2024-04-20"
80          }]
81      }
82  }
```

# References

1. Esteves, B., Pandit, H.J., Rodríguez-Doncel, V.: ODRL profile for expressing consent through granular access control policies in solid. In: 2021 IEEE European Symposium on Security and Privacy Workshops (EuroS PW), pp. 298–306 (2021). https://doi.org/10.1109/EuroSPW54576.2021.00038

2. Esteves, B., Rodríguez-Doncel, V.: Analysis of ontologies and policy languages to represent information flows in GDPR. Semantic Web (Preprint) 1–35 (2022)

3. Iannella, R., Villata, S.: ODRL Information Model 2.2 (2018). https://www.w3.org/TR/odrl-model/

4. ISO/IEC 29100:2011 Information technology—Security techniques—Privacy framework. Technical report, International Standards Organisation (ISO) (2011)

5. ISO/IEC: ISO/IEC 29184:2020 Information technology – Online privacy notices and consent (2020)

6. ISO/IEC: ISO/IEC TS 27560:2023 Privacy technologies—Consent record information structure (2021)

7. Jesus, V., Pandit, H.J.: Consent receipts for a usable and auditable web of personal data. IEEE Access 10, 28545–28563 (2022). https://doi.org/10.1109/ACCESS.2022.3157850

8. Kurteva, A., Chhetri, T.R., Pandit, H.J., Fensel, A.: Consent through the lens of semantics: state of the art survey and best practices. Semantic Web 1–27 (2021)

9. Lizar, M., Turner, D.: Consent Receipt Specification v1.1.0. Technical report, Kantara Initiative (2017)

10. Machuletz, D., Böhme, R.: Multiple purposes, multiple problems: a user study of consent dialogs after GDPR. Proc. Priv. Enhan. Technol. 2020(2), 481–498 (2020)

11. Matte, C., Santos, C., Bielova, N.: Purposes in IAB Europe's TCF: which legal basis and how are they used by advertisers? In: Annual Privacy Forum (APF 2020) (2020)
12. Pandit, H.J., Esteves, B.: Enhancing Data Use Ontology (DUO) for health-data sharing by extending it with ODRL and DPV. Semantic Web 1–26 (2024). https://doi.org/10.3233/SW-243583
13. Pandit, H.J., Esteves, B., Krog, G.P., Ryan, P., Golpayegani, D., Flake, J.: Data privacy vocabulary (dpv)–version 2. In: OSF (2024). arXiv:2404.13426. https://doi.org/10.31219/osf.io/ma9ue
14. Pandit, H.J., Krog, G.P.: Comparison of notice requirements for consent between ISO/IEC 29184: 2020 and general data protection regulation. J. Data Protect. Priv. 4(2), 193–204 (2021)
15. Pandit, H.J., et al.: Creating a vocabulary for data privacy. In: The 18th International Conference on Ontologies, DataBases, and Applications of Semantics (ODBASE2019), Rhodes, Greece, p. 17 (2019)
16. Regulation (EU) 2016/679 of the European Parliament and of the Council of 27 April 2016 on the protection of natural persons with regard to the processing of personal data and on the free movement of such data, and repealing Directive 95/46/EC (General Data Protection Regulation). Official Journal of the European Union **L119** (2016)

# Author Index

M. Jensen et al. (Eds.): APF 2024, LNCS 14831, p. 253, 2024.
https://doi.org/10.1007/978-3-031-68024-3